Lecture Notes in Mathematics 1956

Editors:
J.-M. Morel, Cachan
F. Takens, Groningen
B. Teissier, Paris

Cho-Ho Chu
School of Mathematical Sciences
Queen Mary, University of London
London E1 4NS
United Kingdom
c.chu@qmul.ac.uk

ISBN 978-3-540-69797-8   ISBN 978-3-540-69798-5 (eBook)

DOI 10.1007/978-3-540-69798-5

Lecture Notes in Mathematics ISSN print edition: 0075-8434
                              ISSN electronic edition: 1617-9692

Library of Congress Control Number: 2008930086

Mathematics Subject Classification (2000): 47B38, 47A10, 47D03, 43A85, 17C65, 31C05, 53C35

Cover design: SPi Publishing Services

Printed on acid-free paper

9 8 7 6 5 4 3 2 1

springer.com

Cho-Ho Chu

Matrix Convolution
Operators on Groups

 Springer

To

Clio and Yen

Preface

Recently, the non-associative algebraic analytic structures of the spaces of bounded complex harmonic functions and harmonic functionals, which are eigenfunctions of convolution operators on locally compact groups and their Fourier algebras, have been studied in detail in [13, 14]. It was proposed in [13] to further the investigation in the non-abelian matrix setting which should have wider applications. This research monograph presents some new results and developments in this connection. Indeed, we develop a general theory of matrix convolution operators on L^p spaces of matrix functions on a locally compact group G, for $1 \leq p \leq \infty$, focusing on the spectral properties of these operators and their eigenfunctions, as well as convolution semigroups, and thereby the results in [9, 13, 14] can be subsumed and viewed in perspective in this matrix context. In particular, we describe the L^p-spectrum of these operators and study the algebraic structures of eigenspaces, of which the one corresponding to the largest possible positive eigenvalue is the space of L^p matrix harmonic functions. Of particular interest are the L^∞ matrix harmonic functions which carry the structure of a Jordan triple system. We study contractivity properties of a convolution semigroup of matrix measures and its eigenspaces. Connections with harmonic functions on Riemannian manifolds are discussed.

Some results of this work have been presented in seminars and colloquia in London, Cergy-Pontoise, Hong Kong, Taiwan, Tübingen and York. We thank warmly the audience at these institutions for their inspiration and hospitality, and hope this monograph will also serve as a useful reference for the interested audience.

The author gratefully acknowledges financial support from the University of London Central Research Fund, as well as support of the European Commission through its 6th Framework Programme "Structuring the European Research Area" and the contract RITA-CT-2004-505493, during his visit in 2006 at IHÉS, France, where part of this work was carried out. It is a pleasure to thank several Referees for their generous comments.

Key words and phrases. Matrix-valued measure. Matrix L^p space. Matrix convolution operator. Spectrum and eigenvalue. Matrix harmonic function. Convolution semigroup. Group C*-algebra. JB*-triple. Riemannian symmetric space. Elliptic operator.

Contents

Chapter 1
Introduction

Let G be a locally compact group and $1 \leq p \leq \infty$. In this monograph, we study the basic structures of the convolution operators $f \mapsto f * \sigma$ on L^p spaces of matrix-valued functions on G, induced by a matrix-valued measure σ on G. This study is motivated by recent works in [9, 10, 12–14, 16] on complex and matrix-valued σ-harmonic functions on G which are eigenfunctions of the operator $f \mapsto f * \sigma$, as well as their applications in [45] and the fact that a system of scalar convolution equations is equivalent to a matrix convolution equation. The ubiquity of matrix-valued functions gives another impetus to our investigation, for example, the matrix convolution $f * \sigma$ of a matrix distribution f and a matrix measure σ on $\mathbb{R}^n$ has been used in [49] to study partial differential and convolution equations and recently, applications of vector-valued L^2-convolution operators with matrix-valued kernels have been described in depth in [6], and the Fredholm properties of finite sums of weighted shift operators on ℓ^p spaces of Banach space valued functions on $\mathbb{Z}^n$ have been analysed in detail in [54]. Convolution operators on L^p spaces of real and complex functions are well-studied in literature, however, there are at least two new elements in the matrix setting, namely, the non-commutativity of the matrix multiplication and the non-associative structures of the harmonic functions, which add complexity to the subject and often require more delicate treatment. Some of our results for matrix convolution operators are also new in the scalar case.

Among many well-known examples of convolution operators, the following is relevant to us. Let G be a connected Lie group and let $\mathcal{L}$ be a second order G-invariant elliptic differential operator on G, annihilating the constant functions. Then $\mathcal{L}$ generates a convolution semigroup of probability measures $\{\sigma_t\}_{t>0}$ on G, giving rise to a strongly continuous contractive semigroup $T_t : L^p(G) \longrightarrow L^p(G)$ of convolution operators, where $1 \leq p < \infty$ and

$$T_0 = I, \quad T_t(f) = f * \sigma_t \quad (t > 0).$$

A function $f \in L^\infty(G)$ satisfies $T_t(f) = f$ for all $t > 0$ if, and only if, it is C^2 and $\mathcal{L}$-harmonic on G, that is, $\mathcal{L}f = 0$ (cf. [39] and [1, Proposition V.6]). Moreover, a C^2 L^p-function f on G satisfies $\mathcal{L}f = \alpha f$ if, and only if, $T_t(f) = e^{\alpha t}f$ for all

C.-H. Chu, *Matrix Convolution Operators on Groups*. Lecture Notes in Mathematics 1956, doi: 10.1007/978-3-540-69798-5, © Springer-Verlag Berlin Heidelberg 2008

$t > 0$. A similar example in the matrix setting has been given in [9]. More generally, if $\mathcal{L}$ is a translation invariant Dirichlet form on a locally compact group G, then the semigroup it generates is also a semigroup of convolution operators on $L^p(G)$. In view of these examples, it is natural to include convolution semigroups $\{\sigma_t\}_{t>0}$ of matrix-valued measures in our study. Also, the 1-eigenspace $\{f \in L^\infty(G): f * \sigma_t = f\}$ of T_t, that is, the space of bounded σ_t-*harmonic* functions on G, will be of particular interest to us.

Now we outline the contents of the monograph. Let M_n be the space of $n \times n$ complex matrices. We first introduce, in Chapter 2, L^p spaces of M_n-valued functions on G, denoted by $L^p(G,M_n)$, as a setting for convolution operators. We recall some basic definitions and derive some results for scalar convolution operators in Section 2.1, for later reference. In Section 2.2, we discuss differentiability of the norm in $L^p(G,M_n)$. When M_n is equipped with the Hilbert-Schmidt norm, we compute, in Proposition 2.2.5, the Gateaux derivative of the norm of $L^p(G,M_n)$. This is needed in Chapter 4 for proving some differential inequalities for matrix convolution semigroups in order to derive hypercontractive properties.

We study, in Chapter 3, the matrix convolution operators $T_\sigma : f \in L^p(G,M_n) \mapsto f * \sigma \in L^p(G,M_n)$, where σ is an M_n-valued measure. Due to non-commutativity of the matrix multiplication, we need to introduce the *left* convolution operator $L_\sigma : f \mapsto \sigma *_\ell f$ in order to have a consistent duality theory. In the scalar case, we have $T_\sigma = L_\sigma$. This gives another perspective of the difference between the scalar and matrix cases. We first characterise the matrix convolution operators T_σ, in Section 3.1, and show they are translation invariant operators satisfying some continuity condition. On the matrix L^1 space, they are exactly the operators commuting with left translations. This result is known to be false in the scalar case for L^p spaces if $p \neq 1$, even when G is abelian. We give precise results in the matrix setting for all p. In Section 3.2, we give necessary and sufficient conditions in Theorem 3.2.1 for weak compactness of the convolution operator T_σ on the matrix L^1 and L^∞ spaces. In Section 3.3, we focus on the spectral properties of T_σ. We prove various results concerning its spectrum and eigenvalues. To obtain these results, we introduce the *matrix-valued Fourier transform* and, for abelian groups G, the *determinant* of a matrix-valued measure σ on G. The latter enables us to reduce some arguments to the scalar case. Among other results for abelian groups, we extend the Wiener-Levy theorem to the matrix setting and use it to show in Theorem 3.3.23 that, for an absolutely continuous matrix-valued measure σ on an abelian group G, the L^p-spectrum of T_σ is exactly the closure of the eigenvalues of matrices in the Fourier image of σ. For $p = 2$, absolute continuity of σ is not required and the result follows from a matrix version of the Plancherel theorem for $L^2(G,M_n)$. For non-abelian groups, computation of spectrum is known to be rather complicated and there seem to be fewer definitive results even in the scalar case. Nevertheless, we develop a device to study the L^2 spectrum by identifying the left convolution operator L_σ on $L^2(G,M_n)$ as an element in the tensor product $C_r^*(G) \otimes M_n$ of the reduced group C*-algebra $C_r^*(G)$ and M_n. From this, we are able to deduce several spectral results for T_σ and obtain an extension of the above result for L^2-spectrum to the non-abelian case. We show, in Corollary 3.3.39, that, for absolutely continuous symmetric σ and disregarding 0, the L^2-spectrum of T_σ consists of spectrum of each element in $\hat{\sigma}(\hat{G}_r)$, where $\hat{\sigma}$ is the Fourier transform of σ and $\hat{G}_r$ is the reduced dual of G. However, for compact groups G and absolutely continuous σ, the convolution operator T_σ is compact and the above result for L^p-spectrum of T_σ still holds in this case. As an application, we use the above result for L^2-spectrum to describe the spectrum, in Example 3.3.41, of a discrete Laplacian $\mathcal{L}_d$ of a (possibly infinite) homogeneous graph acted on by a discrete group. If, moreover, $\mathcal{L}_d$ acts on vector-valued functions on the graph, its spectrum is called the *vibrational spectrum* in [20], because of its connection with vibrational modes of molecules, and our result also applies for the case of M_n-valued functions.

The last topic in Chapter 3 concerns the eigenspaces of T_σ:

$$H_\alpha(T_\sigma, L^p(G, M_n)) = \{f \in L^p(G, M_n) : f * \sigma = \alpha f\}.$$

For $\alpha = \|\sigma\|$, which is the largest possible non-negative eigenvalue, the functions in $H_\alpha(T_\sigma, L^p(G, M_n))$ are the M_n-valued L^p σ-*harmonic* functions on G. By normalizing, we consider the space $H_1(T_\sigma, L^p(G, M_n))$ for $\|\sigma\| = 1$ and discuss synthesis for complex-valued harmonic functions on abelian groups. For any group G, we show in Proposition 3.3.56 that there is a contractive projection from $L^p(G, M_n)$ onto $H_1(T_\sigma, L^p(G, M_n))$ and that $H_1(T_\sigma, L^p(G, M_n)) = H_1(L_{\tilde{\sigma}}, L^q(G, M_n))^*$, for $\|\sigma\| = 1$ and $1 < p < \infty$. For $p = \infty$, this result was proved in [9] and it implies that the space $H_1(T_\sigma, L^\infty(G, M_n))$ of bounded σ-harmonic functions carries the structure of a Jordan triple system. The triviality of $H_1(T_\sigma, L^\infty(G, M_n))$, that is, the absence of a non-constant function in $H_1(T_\sigma, L^\infty(G, M_n))$, is a *Liouville type theorem* for σ. Such a Liouville theorem has been proved in [16] when G is nilpotent and σ is positive and non-degenerate. When σ is positive and adapted, it is not difficult to show that $H_1(T_\sigma, L^p(G, M_n))$ is trivial for compact groups G and for all p. For arbitrary groups, we show that $H_1(T_\sigma, L^1(G, M_n))$ has dimension at most n^2.

In Section 3.4, we study Jordan structures of the eigenspace $H_1(T_\sigma, L^\infty(G, M_n))$ and discuss applications to harmonic functions on Riemannian symmetric spaces. To put things in perspective, we first explain how Jordan structures originated from the geometry of Riemannian symmetric spaces. It is therefore interesting that the eigenspace $H_1(T_\sigma, L^\infty(G, M_n))$, which is closely related to harmonic functions on symmetric spaces, also carries a Jordan structure. A symmetric space can be represented as a right coset space G/K of a Lie group G. Furstenberg [29] has characterised bounded harmonic functions on a symmetric space $\Omega = G/K$ of *non-compact type* in terms of convolution of a probability measure σ on G. Making use of this and of our previous results, one can show that the space $H^\infty(\Omega, \mathbb{C})$ of bounded harmonic functions on Ω contains non-constant functions and has the structure of an abelian C*-algebra. This gives a Poisson representation of $H^\infty(\Omega, \mathbb{C})$. We should note that, although this *Jordan C*-approach* is slightly different from [29] and is valid in the wider class of locally compact groups, it is based on the main ideas in [29]. The remaining Section 3.4 is devoted to determining when the space $H_1(T_\sigma, L^\infty(G))$ is a Jordan subtriple of the von Neumann algebra $L^\infty(G)$.

The object of study in Chapter 4 is the convolution semigroup $\mathcal{S} = \{\sigma_t\}_{t>0}$ of matrix-valued measures on G. Our investigation is guided by two objectives. One is the application to harmonic functions on Lie groups. The other concerns contractivity properties of the semigroup. The semigroup $\{\sigma_t\}_{t>0}$ induces a semigroup $\{T_{\sigma_t}\}_{t>0}$ of convolution operators on $L^p(G,M_n)$. Hence our previous results and techniques can be used in this context. For instance, one can show that there is a contractive projection $P : L^p(G,M_n) \longrightarrow L^p(G,M_n)$ with range

$$\bigcap_{t>0} H_1(T_{\sigma_t}, L^p(G,M_n)) = \{f \in L^p(G,M_n) : f = f * \sigma_t \text{ for all } t > 0\}$$

which is the space of matrix L^p harmonic functions for the generator of $\{T_{\sigma_t}\}$, and is denoted by $H_{\mathcal{S}}^p(G,M_n)$. It is a Jordan triple system when $p = \infty$. If $\sigma_t \geq 0$, the triviality of $H_{\mathcal{S}}^\infty(G,M_n)$ implies that G is amenable. If $\{\sigma_t\}_{t>0}$ is generated by the above elliptic operator $\mathcal{L}$ on a connected Lie group G, then $H_{\mathcal{S}}^p(G,\mathbb{C})$ is the space of L^p $\mathcal{L}$-harmonic functions on G and it follows from the spectral theory of T_σ that all L^p $\mathcal{L}$-harmonic functions on G are constant for $1 \leq p < \infty$ (cf. Proposition 4.1.8), and the bounded $\mathcal{L}$-harmonic functions form an abelian C*-algebra which also admits a Poisson representation. The latter result has been proved in [1]. We should remark that the non-existence of a non-constant L^p $\mathcal{L}$-harmonic function on Lie groups, for $1 < p < \infty$, is well-known from a result of Yau [64] for complete Riemannian manifolds. However, an analogous result for $p = 1$ requires non-negativity of the Ricci curvature and cannot be applied directly to Lie groups because Ricci curvature of a Riemannian metric can change sign in Lie groups [50]. In the last section, we extend Gross's result on hypercontractivity for semigroups [36] to the matrix setting, and show in Theorem 4.2.5 that the matrix semigroup $\{T_{\sigma_t}\}_{t>0}$ is hypercontractive if, and only if, its generator satisfies a log-Sobolev type inequality.

Chapter 2
Lebesgue Spaces of Matrix Functions

In this Chapter, we introduce the notations and define the spaces $L^p(G, M_n)$ of matrix L^p functions on locally compact groups G as a setting for later developments. We recall some basic definitions and derive some results for convolution operators in the scalar case. We discuss differentiability of the norm in $L^p(G, M_n)$ which is needed later, and compute the Gateaux derivative of the norm when the matrix space M_n is equipped with the Hilbert-Schmidt norm.

2.1 Preliminaries

We denote by G throughout a locally compact group with identity e and a *right* invariant Haar measure λ. To avoid the inconvenience of additional measure-theoretic technicalities, we assume throughout that λ is σ-finite. If G is compact, λ is normalized to $\lambda(G) = 1$.

Let $1 \leq p < \infty$. Given a complex Banach space E, we denote by $L^p(G, E)$ the Banach space of (equivalence classes of) E-valued Bochner integrable functions f on G satisfying

$$\|f\|_p = \left(\int_G \|f(x)\|^p d\lambda(x) \right)^{\frac{1}{p}} < \infty$$

(cf. [22, p.97]). We write $L^p(G)$ for $L^p(G, E)$ if $\dim E = 1$. In the sequel, E is usually the C^*-algebra M_n of $n \times n$ complex matrices in which case, a function $f : G \longrightarrow M_n$ is an $n \times n$ matrix (f_{ij}) of complex functions f_{ij} on G.

We denote by $\mathcal{B}(E)$ the Banach algebra of bonded linear self-maps on a Banach space E.

Let $\mathrm{Tr} : M_n \to \mathbb{C}$ be the canonical trace of M_n. Every continuous linear functional $\varphi : M_n \to \mathbb{C}$ is of the form $\varphi(\cdot) = \mathrm{Tr}(\cdot A_\varphi)$ where the matrix $A_\varphi \in M_n$ is unique and $\|\varphi\| = \mathrm{Tr}(|A_\varphi|) = \mathrm{Tr}((A_\varphi^* A_\varphi)^{1/2})$ which is the trace-norm $\|A_\varphi\|_{tr}$ of A_φ. We will identify the dual M_n^*, via the map $\varphi \in M_n^* \mapsto A_\varphi \in M_n$, with the vector space M_n equipped with the trace-norm $\| \cdot \|_{tr}$. If we equip M_n with the Hilbert-Schmidt norm

C.-H. Chu, *Matrix Convolution Operators on Groups*. Lecture Notes in Mathematics 1956,
doi: 10.1007/978-3-540-69798-5, © Springer-Verlag Berlin Heidelberg 2008

$\|A\|_{hs} = \mathrm{Tr}(A^*A)^{1/2}$, then M_n is a Hilbert space with inner product $\langle A, B \rangle = \mathrm{Tr}(B^*A)$. We note that the C*-norm, the trace-norm and the Hilbert-Schmidt norm on M_n are related by

$$\|\cdot\| \leq \|\cdot\|_{tr} \leq \sqrt{n}\|\cdot\|_{hs} \leq n\|\cdot\|$$

and norm convergence is equivalent to entry-wise convergence in M_n.

If M_n is equipped with the Hilbert-Schmidt norm, then $L^2(G, (M_n, \|\cdot\|_{hs}))$ is a Hilbert space, with inner product

$$\langle f, g \rangle_2 = \int_G \mathrm{Tr}(f(x)g(x)^*)d\lambda(x).$$

Since $\|f(x)g(x)^*\|_{hs} \leq \|f(x)\|_{hs}\|g(x)\|_{hs}$ for $f, g \in L^2(G, (M_n, \|\cdot\|_{hs}))$, the Bochner integral

$$\langle\langle f, g \rangle\rangle = \int_G f(x)g(x)^* d\lambda(x)$$

exists in M_n and defines an M_n-valued inner product, turning $L^2(G, (M_n, \|\cdot\|_{hs}))$ into an inner product (left) M_n-module.

We denote by $L^\infty(G, M_n)$ the complex Banach space of M_n-valued essentially bounded (locally) λ-measurable functions on G, where M_n is equipped with the C*-norm. It is a von Neumann algebra, with predual $L^1(G, M_n^*)$, under the pointwise product and involution:

$$(fg)(x) = f(x)g(x), \quad f^*(x) = f(x)^* \qquad (f, g \in L^\infty(G, M_n), \ x \in G).$$

We will study convolution operators on $L^p(G, M_n)$ defined by matrix-valued measures. In this section, we first recall some basic definitions and derive some results for convolution operators on $L^p(G)$, for later reference. One important difference in the matrix setting is the presence of non-commutative and non-associative algebraic structures.

We equip the vector space $C(G)$ of complex continuous functions on G with the topology of uniform convergence on compact sets in G, and denote by $C_c(G)$ the subspace of functions with compact support. The Banach space of bounded complex continuous functions on G is denoted by $C_b(G)$. Let $C_0(G)$ be the Banach space of complex continuous functions on G vanishing at infinity. The dual $C_0(G)^*$ identifies with the space $M(G)$ of complex regular Borel measures on G. Each $\mu \in M(G)$ has finite total variation $|\mu|$ and $M(G)$ is a unital Banach algebra in the total variation norm and the convolution product:

$$\|\mu\| = |\mu|(G), \quad \langle f, \mu * \nu \rangle = \int_G \int_G f(xy)d\mu(x)d\nu(y) \qquad (f \in C_0(G), \mu, \nu \in M(G))$$

where we always denote the duality of a dual pair of Banach spaces E and F by

$$\langle \cdot, \cdot \rangle : E \times F \longrightarrow \mathbb{C}.$$

We also write $\mu(f)$ for $\langle f, \mu \rangle = \int_G f d\mu$. The unit mass at a point $a \in G$ is denoted by δ_a where δ_e is the identity in $M(G)$. A measure $\mu \in M(G)$ is called *absolutely continuous* if its total variation $|\mu|$ is absolutely continuous with respect to the Haar measure λ.

Given $\sigma \in M(G)$, the *support* of σ is defined to be the support of its total variation $|\sigma|$ and is denoted by supp σ. We denote by G_σ the closed subgroup of G generated by the support of $|\sigma|$. A measure $\sigma \in M(G)$ is called *adapted* if $G_\sigma = G$. A measure $\sigma \in M(G)$ is said to be *non-degenerate* if supp $|\sigma|$ generates a dense semigroup in G. Evidently, every non-degenerate measure is adapted. An absolutely continuous (non-zero) measure on a *connected* group must be adapted.

By a *(complex) measure* μ on G, we will mean a measure $\mu \in M(G) \backslash \{0\}$.

The convolutions for Borel functions f and g on G, when exit, are defined by

$$(f * g)(x) = \int_G f(xy^{-1})g(y)d\lambda(y);$$

$$(f * \mu)(x) = \int_G f(xy^{-1})d\mu(y);$$

$$(\mu * f)(x) = \int_G f(y^{-1}x)\triangle_G(y^{-1})d\mu(y)$$

where $\triangle_G$ is the modular function satisfying $d\lambda(xy) = \triangle_G(x)d\lambda(y)$ and $d\lambda(x^{-1}) = \triangle_G(x^{-1})d\lambda(x)$.

We denote by ℓ_x and r_x, respectively, the left and right translations by an element $x \in G$:

$$\ell_x f(y) = f(x^{-1}y), \qquad r_x f(y) = f(yx) \qquad (y \in G)$$

for any function f on G. A complex function f on G is left uniformly continuous if $\|r_x f - f\|_\infty \longrightarrow 0$ as $x \to e$. It is right uniformly continuous if $\|\ell_x f - f\|_\infty \longrightarrow 0$ as $x \to e$. We also write $_x f = \ell_{x^{-1}} f$ and f_x for $r_x f$.

We note that each $f \in C_c(G)$ is both left and right uniformly continuous, and for any $\mu \in M(G)$, we have $f * \mu \in C_b(G)$ since $|f * \mu(x) - f * \mu(y)| \leq \|\ell_{xy^{-1}} f - f\| \|\mu\|$. We also have

$$\langle f, \mu * \nu \rangle = \langle \tilde{f}, \tilde{\nu} * \tilde{\mu} \rangle \tag{2.1}$$

where $\nu \in M(G)$ and we define $\tilde{f}(x) = f(x^{-1})$ and $d\tilde{\mu}(x) = d\mu(x^{-1})$. Note that

$$\tilde{\mu}(f) = \mu(\tilde{f}) = (f * \mu)(e) \quad \text{and} \quad \widetilde{\mu * \nu} = \tilde{\nu} * \tilde{\mu}$$

for $f \in C_c(G)$.

Let $\sigma \in M(G)$. For $1 \leq p \leq \infty$, we define the convolution operator $T_\sigma : L^p(G) \longrightarrow L^p(G)$ by

$$T_\sigma(f) = f * \sigma \qquad (f \in L^p(G)).$$

To avoid triviality, σ is always non-zero for T_σ. The definition of T_σ depends on its domain $L^p(G)$ although we often omit referring to it if there is no ambiguity. When regarded as an operator on $L^p(G)$, the operator T_σ is easily seen to be bounded and we denote its norm by $\|T_\sigma\|_p$, or simply $\|T_\sigma\|$ in obvious context. We have $\|T_\sigma\|_p \leq \|\sigma\|$.

A convolution operator $T_\sigma : L^p(G) \longrightarrow L^p(G)$ commutes with left translations:

$$\ell_x T_\sigma = T_\sigma \ell_x \qquad (x \in G).$$

Conversely, for abelian groups G, every translation invariant operator $T : L^1(G) \longrightarrow L^1(G)$ is a convolution operator T_σ for some $\sigma \in M(G)$ [55, 3.8.4]. However, this result does not hold for $1 < p \leq \infty$, even if G is compact and abelian [44, p.85]. We will characterise the more general matrix convolution operators in Chapter 3. In particular, the above L^1 result is generalized to the matrix-valued case, for all locally compact groups.

For $1 \leq p \leq \infty$, we denote by q its conjugate exponent throughout, that is, $\frac{1}{p} + \frac{1}{q} = 1$, and for the dual pairing $\langle \cdot, \cdot \rangle$ between $L^p(G)$ and $L^q(G)$, we have

$$\langle f * \sigma, h \rangle = \langle f, h * \widetilde{\sigma} \rangle \tag{2.2}$$

for $f \in L^p(G)$ and $h \in L^q(G)$. This implies that T_σ is weakly continuous on $L^p(G)$ for $1 \leq p < \infty$, and is weak* continuous on $L^\infty(G)$. In particular, T_σ is a weakly compact operator on $L^p(G)$ for $1 < p < \infty$. For $p = 1, \infty$, we will discuss presently weak compactness of $T_\sigma : L^p(G) \longrightarrow L^p(G)$, but we note the following two lemmas first.

Lemma 2.1.1. *Let $\sigma \in M(G)$ and $p < \infty$. Let $T_\sigma^* : L^q(G) \longrightarrow L^q(G)$ be the dual map of the convolution operator $T_\sigma : L^p(G) \longrightarrow L^p(G)$. Then $T_\sigma^* = T_{\widetilde{\sigma}}$. The operator $T_\sigma : L^2(G) \longrightarrow L^2(G)$ is self-adjoint if $\widetilde{\sigma} = \sigma$ is a real measure. The weak* continuous operator $T_\sigma : L^\infty(G) \longrightarrow L^\infty(G)$ has predual $T_{\widetilde{\sigma}} : L^1(G) \longrightarrow L^1(G)$.*

Proof. By (2.2), we have $\langle f, T_\sigma^* h \rangle = \langle f, T_{\widetilde{\sigma}} h \rangle$ for $f \in L^p(G)$ and $h \in L^q(G)$. The adjoint of T_σ in $\mathcal{B}(L^2(G))$ is $T_{\overline{\widetilde{\sigma}}}$ where $\overline{\sigma}$ is the complex conjugate of σ. $\qquad\square$

Lemma 2.1.2. *Let $\sigma \in M(G)$ and let T_σ be the convolution operator on $L^p(G)$ for $p = 1, \infty$. We have $\|T_\sigma\|_1 = \|T_\sigma\|_\infty = \|\sigma\|$.*

Proof. We have $\|\sigma\| = \sup\{|\int_G f d\sigma| : f \in C_c(G) \text{ and } \|f\| \leq 1\}$ in which

$$\left| \int_G f d\sigma \right| = |\widetilde{f} * \sigma(e)| \leq \|\widetilde{f} * \sigma\|_\infty \leq \|T_\sigma\|_\infty$$

where $\widetilde{f} * \sigma \in C_b(G)$. Next, we have $\|T_\sigma\|_1 = \|T_\sigma^*\|_\infty = \|T_{\widetilde{\sigma}}\|_\infty = \|\widetilde{\sigma}\| = \|\sigma\|$. $\qquad\square$

Remark 2.1.3. We note that $\|T_\sigma\|_p$ need not equal $\|\sigma\|$ if $1 < p < \infty$. Indeed, if σ is an adapted probability measure whose support contains the identity e and if $\|T_\sigma\|_p = 1$ for some $1 < p < \infty$, then G is amenable (see, for example, [4, Theorem 1]). On the other hand, if G is amenable and σ is a probability measure, then $\|T_\sigma\|_p = 1$ for all p (cf. [33, p.48]).

By Lemma 2.1.2, the spectral radius of $T_\sigma \in \mathcal{B}(L^p(G))$, for $p = 1, \infty$, is $\lim_n \|T_\sigma^n\|^{\frac{1}{n}} = \lim_n \|T_{\sigma^n}\|^{\frac{1}{n}} = \lim_n \|\sigma^n\|^{\frac{1}{n}}$ where σ^n is the n-fold convolution of σ with itself.

Lemma 2.1.4. *Let G be a compact group and let $\sigma \in M(G)$ be absolutely contin-uous. Then the convolution operator $T_\sigma : L^p(G) \longrightarrow L^p(G)$ is compact for every $p \in [1, \infty]$.*

Proof. Let $\sigma = h \cdot \lambda$ for some $h \in L^1(G)$. Consider first $T_\sigma : L^\infty(G) \longrightarrow L^\infty(G)$. By absolute continuity of σ, we have $T_\sigma(L^\infty(G)) \subset C(G)$. Hence, by Arzela-Ascoli theorem, we need only show that the set

$$\{T_\sigma(f) : \|f\|_\infty \leq 1\}$$

is equicontinuous in $C(G)$. Let $\varepsilon > 0$. Pick $\varphi \in C_c(G)$ with support K and $\|\varphi - h\|_1 < \frac{\varepsilon}{4}$. Let W be a compact neighbourhood of the identity $e \in G$. By uniform continuity, we can choose a compact neighbourhood $V \subset W$ of e such that

$$|\varphi(x) - \varphi(y)| < \frac{\varepsilon}{2\lambda(KW)}$$

whenever $x^{-1}y \in V$. Then

$$\|\varphi_x - \varphi_y\|_1 = \int_G |\varphi(zx) - \varphi(zy)| d\lambda(z)$$
$$= \int_{KW} |\varphi(z) - \varphi(zx^{-1}y)| d\lambda(z) < \frac{\varepsilon}{2}.$$

It follows that, for $x^{-1}y \in V$ and $\|f\|_\infty \leq 1$, we have

$$|T_\sigma(f)(x) - T_\sigma(f)(y)| = \left| \int_G f(xz^{-1})h(z)d\lambda(z) - \int_G f(yz^{-1})h(z)d\lambda(z) \right|$$
$$\leq \int_G |f(z^{-1})h(zx) - f(z^{-1})h(zy)| d\lambda(z)$$
$$\leq \|f\|_\infty \|h_x - h_y\|_1$$
$$\leq \|f\|_\infty (\|h_x - \varphi_x\|_1 + \|\varphi_x - \varphi_y\|_1 + \|h_y - \varphi_y\|_1) < \varepsilon$$

which proves equicontinuity and hence, compactness of $T_\sigma : L^\infty(G) \longrightarrow L^\infty(G)$.

Likewise $T_{\tilde{\sigma}} : L^\infty(G) \longrightarrow L^\infty(G)$ is compact and therefore $T_\sigma : L^1(G) \longrightarrow L^1(G)$ is compact.

Let $1 < p < \infty$. Let (h_n) be a sequence in $C(G)$ such that $\|h_n - h\|_1 \longrightarrow 0$. Then $T_\sigma = \lim_{n \to \infty} T_{\sigma_n}$ in $\mathcal{B}(L^p(G))$, where $\sigma_n = h_n \cdot \lambda$. Hence it suffices to show compactness of T_σ on $L^p(G)$ for the case $h \in C(G)$.

Let (f_n) be a sequence in the unit ball of $L^p(G)$. Then $\|f_n\|_1 \leq 1$ for all n and compactness of $T_\sigma : L^1(G) \longrightarrow L^1(G)$ implies that the sequence $(f_n * \sigma)$ contains a subsequence L^1-converging to some $f \in L^1(G)$, and hence a subsequence $(f_k * \sigma)$ converging pointwise to f λ-almost everywhere. Since $h \in C(G)$, we have $\|f_k * \sigma\|_\infty \leq \|f_k\|_p \|h\|_q \leq \|h\|_q$ for all k, and $f \in L^\infty(G)$. It follows that

$$\|f_k * \sigma - f\|_p^p \le \|f_k * \sigma - f\|_1 \|f_k * \sigma - f\|_\infty^{p-1} \longrightarrow 0 \qquad \text{as } k \to \infty.$$

This proves compactness of $T_\sigma : L^p(G) \longrightarrow L^p(G)$. $\square$

Remark 2.1.5. The above result is clearly false if σ is not absolute continuous, for instance, T_σ is the identity operator if $\sigma = \delta_e$.

A compactness criterion has been given in [48] for a class of convolution operators of the form $f \in L^1(G) \mapsto f * F \in C(G)$ where $F \in L^\infty(G)$ and G is compact abelian. Compactness of the composition of a convolution operator with a multiplier has also been considered in [59, 60]. Fredholmness of convolution operators on locally compact groups has been studied in [54, 59, 61].

Proposition 2.1.6. *Let σ be a positive measure on a group G such that $\sigma^2 * \widetilde{\sigma}^2$ is adapted. Let T_σ be the associated convolution operator. The following conditions are equivalent.*

  (i) $T_\sigma : L^1(G) \longrightarrow L^1(G)$ *is weakly compact.*
  (ii) $T_\sigma : L^1(G) \longrightarrow L^1(G)$ *is compact.*
  (iii) $T_\sigma : L^\infty(G) \longrightarrow L^\infty(G)$ *is weakly compact.*
  (iv) $T_\sigma : L^\infty(G) \longrightarrow L^\infty(G)$ *is compact.*
  (v) $T_\sigma : L^p(G) \longrightarrow L^p(G)$ *is compact for all $p \in [1, \infty]$.*
  (vi) G *is compact and σ is absolutely continuous.*

Proof. (i) $\Longrightarrow$ (vi). We first prove compactness of G. Note that $L^1(G)$ has the Dunford-Pettis property and in particular, every weakly compact operator on $L^1(G)$ sends weakly compact subsets to norm compact sets [22, p.154]. Hence weak compactness of T_σ implies that the operator $T_\sigma^2 : L^1(G) \longrightarrow L^1(G)$ is compact, and so is the operator $T_{\sigma*\sigma*\widetilde{\sigma}*\widetilde{\sigma}} = T_\sigma^2 T_{\widetilde{\sigma}}^2$. Since $\sigma^2 * \widetilde{\sigma}^2$ is a positive measure, the spectral radius of $T_{\sigma^2*\widetilde{\sigma}^2}$ is $\sigma(G)^4$, by a remark before Lemma 2.1.4. On the Hilbert space $L^2(G)$, the operator $T_{\sigma^2*\widetilde{\sigma}^2} = T_{\sigma^2}^* T_{\sigma^2}$ is a positive operator and therefore has only non-negative eigenvalues. The eigenvalues of $T_{\sigma^2*\widetilde{\sigma}^2} \in \mathcal{B}(L^1(G))$ are also eigenvalues of $T_{\sigma^2*\widetilde{\sigma}^2} \in \mathcal{B}(L^2(G))$ and therefore non-negative. It follows that $\sigma(G)^4$ is an eigenvalue of the compact operator $T_{\sigma^2*\widetilde{\sigma}^2} \in \mathcal{B}(L^1(G))$, that is, there is a non-zero function $f \in L^1(G)$ satisfying $f * \sigma^2 * \widetilde{\sigma}^2 = \sigma(G)^4 f$. Note that the measure $\sigma(G)^{-4}\sigma^2 * \widetilde{\sigma}^2$ is an adapted probability measure on G. Now, by [10, Theorem 3.12], f is constant which implies that G must be compact.

Next, we show that σ is absolutely continuous. By the Dunford-Pettis-Phillips Theorem [22, p.75], there is an essentially bounded function $g : G \longrightarrow L^1(G)$ such that

$$T_\sigma(f) = \int_G fg d\lambda \qquad (f \in L^1(G)).$$

Now the arguments in [22, p.91] can still be applied without commutativity of G. Let $a \in G$. For each $f \in L^1(G)$, we have, for λ-a.e. y,

$$\int_G f(x)g(x)(y)d\lambda(x) = T_\sigma f(y) = \ell_{a^{-1}}T_\sigma(\ell_a f)(y)$$
$$= \int_G (\ell_a f)(x)g(x)(ay)d\lambda(x)$$
$$= \int_G f(a^{-1}x)g(x)(ay)d\lambda(x)$$
$$= \int_G f(x)g(ax)(ay)d\lambda(x).$$

It follows that

$$g(ax)(ay) = g(x)(y)$$

for λ-a.e. x and y. This implies that, for each $f \in C(G)$, the function

$$F(y) = \int_G f(x)g(yx^{-1})(y)d\lambda(x) \qquad (y \in G)$$

is invariant under the left translations ℓ_a for all $a \in G$. Using compactness of G, one can show that F is constant λ-almost every on G, as in [22, p.91], and hence we have,

$$F(y) = \int_G F(z)d\lambda(z) = \int_G \int_G f(x)g(zx^{-1})(x)d\lambda(x)d\lambda(z) \qquad (2.3)$$

for λ-a.e. y. Let $h \in L^1(G)$ be defined by

$$h(x) = \int_G g(yx^{-1})(y)d\lambda(y).$$

We show that $f * \sigma = f * h$ for each $f \in C(G) \subset L^1(G)$ which then yields absolutely continuity of σ. Indeed, for each $k \in L^\infty(G)$, we have

$$\langle k, f*h \rangle = \int_G k(y) \int_G f(yx^{-1})h(x)d\lambda(x)d\lambda(y)$$
$$= \int_G k(y) \int_G \int_G f(yx^{-1})g(zx^{-1})(z)d\lambda(x)d\lambda(z)d\lambda(y)$$
$$= \int_G k(y) \int_G f(yx^{-1})g(yx^{-1})(y)d\lambda(x)d\lambda(y) \qquad \text{(by (2.3))}$$
$$= \int_G k(y)f*\sigma(y)d\lambda(y) = \langle k, f*\sigma \rangle$$

which concludes the proof.

(vi) $\implies$ (v). By Lemma 2.1.4.

(v) $\implies$ (iv) $\implies$ (iii). Trivial.

(iii) $\implies$ (ii). The given condition implies that $T_{\tilde\sigma} : L^1 G) \longrightarrow L^1(G)$ is weakly compact. Repeating (i) $\implies$ (v) $\implies$ (iv) for $\tilde\sigma$, we see that $T_{\tilde\sigma} : L^\infty(G) \longrightarrow L^\infty(G)$ is compact, and hence $T_\sigma : L^1(G) \longrightarrow L^1(G)$ is compact.

(ii) $\implies$ (i). Trivial. $\qquad\qquad\qquad\qquad\qquad\qquad\qquad\qquad\qquad\qquad\qquad$ □

Remark 2.1.7. In (i) $\Longrightarrow$ (vi) above, the proof of absolute continuity of σ from weak compactness of $T_\sigma \in B(L^1(G))$ is valid for *any* measure σ on a compact group G, without adaptedness of $\sigma^2 * \widetilde{\sigma}^2$.

Corollary 2.1.8. *Given a positive absolutely continuous measure σ on a connected group G, the following conditions are equivalent.*

(i) $T_\sigma : L^1(G) \longrightarrow L^1(G)$ *is weakly compact.*
(ii) $T_\sigma : L^\infty(G) \longrightarrow L^\infty(G)$ *is weakly compact.*
(iii) $T_\sigma : L^p(G) \longrightarrow L^p(G)$ *is compact for all $p \in [1,\infty]$.*
(iv) G *is compact.*

Proof. This is because absolutely continuous measures on a connected group are adapted.   □

Definition 2.1.9. The spectrum of an element a in a unital Banach algebra $\mathcal{A}$ is denoted by $\operatorname{Spec}_{\mathcal{A}} a$ which is often shortened to $\operatorname{Spec} a$ if the Banach algebra $\mathcal{A}$ is understood. For $1 \le p \le \infty$, we write $\operatorname{Spec}(T_\sigma, L^p(G))$, or simply, $\operatorname{Spec}(T_\sigma, L^p)$, for the spectrum $\operatorname{Spec} T_\sigma$, when regarding $T_\sigma \in \mathcal{B}(L^p(G))$. We denote by $\Lambda(T_\sigma, L^p(G))$, or simply, $\Lambda(T_\sigma, L^p)$, the set of eigenvalues of $T_\sigma : L^p(G) \longrightarrow L^p(G)$.

Given any Banach algebra $\mathcal{A}$ and an element $a \in \mathcal{A}$, we define, as usual, the *quasi-spectrum* of a, denoted by $\operatorname{Spec}'_{\mathcal{A}} a$, to be the spectrum $\operatorname{Spec}_{\mathcal{A}_1} a$ of a in the unit extension $\mathcal{A}_1$ of $\mathcal{A}$. We always have $0 \in \operatorname{Spec}'_{\mathcal{A}} a$. If $\mathcal{A}$ has an identity, then we have

$$\operatorname{Spec}'_{\mathcal{A}} a = \operatorname{Spec}_{\mathcal{A}} a \cup \{0\}.$$

We recall that

$$\operatorname{Spec}(T_\sigma, L^p) = \Lambda(T_\sigma, L^p) \cup \operatorname{Spec}^r(T_\sigma, L^p) \cup \operatorname{Spec}^c(T_\sigma, L^p)$$

where $\operatorname{Spec}^r(T_\sigma, L^p)$ denotes the *residue spectrum* of T_σ, consisting of $\alpha \in \operatorname{Spec}(T_\sigma, L^p) \backslash \Lambda(T_\sigma, L^p)$ satisfying

$$\overline{(T_\sigma - \alpha I)(L^p(G))} \ne L^p(G)$$

and $\operatorname{Spec}^c(T_\sigma, L^p)$ denotes the *continuous spectrum* of T_σ, consisting of $\alpha \in \operatorname{Spec}(T_\sigma, L^p) \backslash \Lambda(T_\sigma, L^p)$ such that

$$\overline{(T_\sigma - \alpha I)(L^p(G))} = L^p(G).$$

Since $T_\sigma^* = T_{\widetilde{\sigma}}$ for $p < \infty$, we have

$$\operatorname{Spec}(T_\sigma, L^p) = \operatorname{Spec}(T_{\widetilde{\sigma}}, L^q)$$

for $1 \le p < \infty$, and also $\operatorname{Spec}(T_\sigma, L^\infty) = \operatorname{Spec}(T_{\widetilde{\sigma}}, L^1)$.

We denote by $\operatorname{Spec} \sigma$ the spectrum of σ in the measure algebra $M(G)$. Note that $\operatorname{Spec} \sigma = \operatorname{Spec} \widetilde{\sigma}$ since $\widetilde{\sigma} * \widetilde{\mu} = \widetilde{\mu * \sigma}$ for each $\mu \in M(G)$.

Given a locally compact group G, we let $\widehat{G}$ be the dual space consisting of (the equivalence classes of) continuous unitary irreducible representations $\pi : G \longrightarrow \mathcal{B}(H_\pi)$, where H_π is a Hilbert space. Let $\iota \in \widehat{G}$ be the one-dimensional identity representation. For $\pi \in \widehat{G}$, $\sigma \in M(G)$ and $f \in L^1(G)$, we define the *Fourier transforms*:

$$\widehat{\sigma}(\pi) = \int_G \pi(x^{-1})d\sigma(x) \in \mathcal{B}(H_\pi),$$

$$\widehat{f}(\pi) = \int_G f(x)\pi(x^{-1})d\lambda(x) \in \mathcal{B}(H_\pi).$$

We have $\widehat{f * \sigma}(\pi) = \widehat{\sigma}(\pi)\widehat{f}(\pi)$ and $\widehat{\mu * \sigma}(\pi) = \widehat{\sigma}(\pi)\widehat{\mu}(\pi)$ for $\mu \in M(G)$.

The spectrum $\mathrm{Spec}_{\mathcal{B}(H_\pi)}\,\widehat{\sigma}(\pi)$ of $\widehat{\sigma}(\pi) \in \mathcal{B}(H_\pi)$ will be written as $\mathrm{Spec}\,\widehat{\sigma}(\pi)$ if no confusion is likely.

If G is abelian, $\widehat{G}$ is the group of characters and we often use the letter χ to denote an element in $\widehat{G}$. For $1 < p < 2$ and $f \in L^p(G)$, we define the Fourier transform $\widehat{f} \in L^q(\widehat{G})$ via Riesz-Thorin interpolation.

A continuous homomorphism χ from an abelian group G to the multiplicative group $\mathbb{C}\backslash\{0\}$ is called a *generalized character*. For such a character χ with $|\chi(\cdot)| \leq 1$, one can still define $\widehat{\sigma}(\chi)$ as above. The spectrum $\Omega(G)$ of the Banach algebra $M(G)$, i.e., the non-zero multiplicative functionals on $M(G)$, identifies with the generalized characters χ of G with $|\chi(\cdot)| \leq 1$, and by Gelfand theory, we have $\mathrm{Spec}\,\sigma = \widehat{\sigma}(\Omega(G))$ which contains $\widehat{\sigma}(\widehat{G})$. The spectrum of $L^1(G)$ identifies with the dual group $\widehat{G}$ and if G is discrete, then $M(G) = \ell^1(G)$ and $\mathrm{Spec}\,\sigma = \mathrm{Spec}_{\ell^1(G)}\,\sigma = \widehat{\sigma}(\widehat{G})$. For arbitrary groups, we have the following result.

Lemma 2.1.10. *Let σ be a complex measure on a group G. Then*

$$\Lambda(T_\sigma, L^1) \subset \bigcup_{\pi \in \widehat{G}} \mathrm{Spec}\,\widehat{\sigma}(\pi) \subset \mathrm{Spec}\,\sigma.$$

The inclusions are strict in general.

Proof. Similar inclusions hold in the more general matrix setting for which a simple proof will be given in Proposition 3.3.8. If σ is an adapted probability measure and G is non-compact, then by [10, Theorem 3.12], $1 \notin \Lambda(T_\sigma, L^1)$ while $1 \in \mathrm{Spec}\,\iota(\sigma)$ where $\iota \in \widehat{G}$ is the identity representation.

If G is abelian, then $\widehat{\sigma}(\widehat{G}) = \bigcup_{\pi \in \widehat{G}} \mathrm{Spec}\,\widehat{\sigma}(\pi)$ and Example 3.3.4 shows that the last inclusion can be strict. In fact, even the closure $\overline{\widehat{\sigma}(\widehat{G})}$ may not equal $\mathrm{Spec}\,\sigma$ by Remark 3.3.24.   □

It has been shown in [10, Lemma 3.11] that $1 \notin \bigcup_{\pi \in \widehat{G}\backslash\{\iota\}} \mathrm{Spec}\,\widehat{\sigma}(\pi)$ if σ is an adapted probability measure on a locally compact group G. In general, there seem to be few definitive results concerning the spectrum of T_σ for non-abelian groups. We will consider this case in Chapter 3 and prove various results there.

We will make use of a version of the Wiener-Levy theorem, stated below, which has been proved in [55, Theorem 6.2.4] and will be generalized to the matrix setting in Chapter 3.

Lemma 2.1.11. *Let Ω be an open set in $\mathbb{C}$ and let $F : \Omega \longrightarrow \mathbb{C}$ be a real analytic function satisfying $F(0) = 0$ if $0 \in \Omega$. Given an abelian group G and a function $f \in L^1(G)$ such that $\overline{\widehat{f}(\widehat{G})} \subset \Omega$, then $F(\widehat{f})$ is the Fourier transform of an $L^1(G)$-function.*

Example 2.1.12. For the Cauchy distribution

$$d\sigma_t(x) = \frac{t}{\pi(t^2 + x^2)}dx \qquad (t > 0)$$

on $\mathbb{R}$, we have $\widehat{\sigma}_t(\widehat{\mathbb{R}}) = \{\exp(-t|x|) : x \in \mathbb{R}\} = (0,1] = \operatorname{Spec}(T_\sigma, L^p) \setminus \{0\} = \Lambda(T_{\sigma_t}, L^\infty)$.

Example 2.1.13. Let G be any locally compact group and let $\sigma = \delta_a$ be the unit mass at $a \in G$. Then T_σ is a translation on $L^p(G)$ and we have

$$\operatorname{Spec}(T_\sigma, L^\infty) \subset \{\alpha : |\alpha| = 1\}.$$

If $G = \mathbb{T}$ and $a = i$, then $L^\infty(\mathbb{T}) \subset L^2(\mathbb{T})$ and $\operatorname{Spec}(T_\sigma, L^\infty) = \operatorname{Spec}(T_\sigma, L^2) = \widehat{\sigma}(\mathbb{Z}) = \{\exp(-in\pi/2) : n \in \mathbb{Z}\} = \{\pm 1, \pm i\} \neq \{\alpha : |\alpha| = 1\}$.

If $G = \mathbb{Z}$ and $a = 1$, then $\operatorname{Spec}(T_\sigma, \ell^2) = \{\alpha : |\alpha| = 1\} = \operatorname{Spec}(T_\sigma, \ell^\infty)$.

If $G = \mathbb{R}$ and $a \neq 0$, then $\operatorname{Spec}(T_\sigma, L^p) = \{\alpha : |\alpha| = 1\} = \Lambda(T_\sigma, L^\infty)$ as $\widehat{\delta}_a(\widehat{\mathbb{R}}) = \{\exp(-ia\theta) : \theta \in \mathbb{R}\}$.

Next consider the measure $\mu = \dfrac{1}{2}(\delta_0 + \delta_1)$ on $\mathbb{R}$. Its n-fold convolution

$$\mu^n = \frac{1}{2^n} \sum_{k=0}^{n} \binom{n}{k} \delta_k$$

is a convex sum of discrete measures and we have

$$\operatorname{Spec}(T_{\mu^n}, L^\infty) = \Lambda(T_{\mu^n}, L^\infty) = \left\{ \frac{1}{2^n} \sum_{k=0}^{n} \binom{n}{k} \exp(-ik\theta) : \theta \in \mathbb{R} \right\}$$

where, for example, $\operatorname{Spec}(T_\mu, L^\infty)$ is the circle containing 0 and internally tangent to the unit circle at 1, with $\sin \pi x$ as a 0-eigenfunction.

2.2 Differentiability of Norm in $L^p(G, M_n)$

We will be working with the complex Lebesgue spaces $L^p(G, M_n)$ where, for convenience and consistency with previous and related works elsewhere, we equip M_n with the C*-norm unless otherwise stated. Some remarks are in order here. First, there is no essential difference if one chooses to equip M_n with the trace norm since it amounts to considering the space $L^p(G, M_n^*)$ which is, for $p > 1$, the dual of $L^q(G, M_n)$. Also, the Lebesgue spaces $L^p(G, M_n)$ defined in terms of the C*, trace and Hilbert-Schmidt norms on M_n are all isomorphic and most results for these three cases are identical. There is, however, a difference among the three cases if one considers the differentiability of the norm of $L^p(G, M_n)$ which will be needed later.

Let us first consider the differentiability of the C*-norm $\| \cdot \|$, the trace norm $\| \cdot \|_{tr}$ and the Hilbert-Schmidt norm $\| \cdot \|_{hs}$ on M_n, regarded as a real Banach space.

We recall that the norm $\| \cdot \|$ of a real Banach space E is said to be *Gateaux differentiable* at a point $u \in E$ if the following limit exists

$$\partial \|u\|(x) = \lim_{t \to 0} \frac{\|u + tx\| - \|u\|}{t}$$

for each $x \in E$, in which case, the limit is called the *Gateaux derivative* of the norm at u, *in the direction of x*. We note that the *right directional derivative*

$$\partial^+ \|u\|(x) = \lim_{t \downarrow 0} \frac{\|u + tx\| - \|u\|}{t}$$

always exists. In fact, it is equal to

$$\sup\{\psi(x) : \psi \text{ is a subdifferential at } u\}$$

where a linear functional ψ in the dual E^* is called a *subdifferential* at u if

$$\psi(x - u) \le \|x\| - \|u\|$$

for each $x \in E$. The norm is Gateaux differentiable at u if, and only if, there is a unique subdifferential at u, in which case, the subdifferential is the Gateaux derivative (cf. [53, Proposition 1.8]).

The Hilbert-Schmidt norm $\| \cdot \|_{hs}$ on M_n is Gateaux differentiable at every $A \in M_n \backslash \{0\}$. Indeed, we have

$$\lim_{t \to 0} \frac{\|A + tX\|_{hs} - \|A\|_{hs}}{t} = \lim_{t \to 0} \frac{\text{Tr}((A + tX)^*(A + tX)) - \text{Tr}(A^*A)}{t(\|A + tX\|_{hs} + \|A\|_{hs})}$$

$$= \frac{\text{Tr}(A^*X + X^*A)}{2\|A\|_{hs}}$$

$$= \frac{1}{\|A\|_{hs}} \text{Re} \, \text{Tr}(A^*X).$$

Although the norm of a separable Banach space is Gateaux differentiable on a dense G_δ set, it is easy to see that the C*-norm and the trace norm need not be Gateaux differentiable at every non-zero $A \in M_n$.

Lemma 2.2.1. *Let $A \in M_n \backslash \{0\}$. The C*-norm on M_n is Gateaux differentiable at A if, and only if, given any unit vectors $\xi, \eta \in \mathbb{C}^n$ with $\|A\xi\| = \|A\eta\| = \|A\|$, we have*

$$\langle A\xi, X\xi \rangle = \langle A\eta, X\eta \rangle \qquad (X \in M_n).$$

In the above case, the Gateaux derivative at A is given by

$$\partial \|A\|(X) = \frac{1}{\|A\|} \mathrm{Re}\, \langle A\xi, X\xi \rangle \qquad (X \in M_n)$$

where $\xi \in \mathbb{C}^n$ is a unit vector satisfying $\|A\xi\| = \|A\|$.

Proof. Suppose the norm is Gateaux differentiable at A. Let $\xi \in \mathbb{C}^n$ be a unit vector such that $\|A\| = \|A\xi\|$. Define a real continuous linear functional $\psi_\xi : M_n \longrightarrow \mathbb{R}$ by

$$\psi_\xi(X) = \frac{1}{\|A\|} \mathrm{Re}\, \langle A\xi, X\xi \rangle \qquad (X \in M_n).$$

Then for each $X \in M_n$, we have

$$
\begin{aligned}
\psi_\xi(X - A) &= \frac{1}{\|A\|} \mathrm{Re}\, \langle A\xi, X - A\xi \rangle \\
&= \frac{1}{\|A\|} \mathrm{Re}\, (\langle A\xi, X\xi \rangle - \langle A\xi, A\xi \rangle) \le \|X\| - \|A\|.
\end{aligned}
$$

Hence ψ_ξ is a subdifferential at A. If η is a unit vector in $\mathbb{C}^n$ such that $\|A\eta\| = \|A\|$, then we must have $\psi_\eta = \psi_\xi$, by uniqueness of the subdifferential, which gives $\langle A\xi, X\xi \rangle = \langle A\eta, X\eta \rangle$ for every $X \in M_n$.

To show the converse, we note that (cf. [5, Proposition 4.12]),

$$\lim_{t \downarrow 0} \frac{\|A + tX\| - \|A\|}{t} = \sup \left\{ \lim_{t \downarrow 0} \frac{\|(A + tX)\xi\| - \|A\xi\|}{t} : \|\xi\| = 1, \|A\xi\| = \|A\| \right\}$$

where

$$
\begin{aligned}
\lim_{t \downarrow 0} \frac{\|(A + tX)\xi\| - \|A\xi\|}{t} &= \lim_{t \downarrow 0} \frac{\langle (A + tX)\xi, (A + tX)\xi \rangle - \langle A\xi, A\xi \rangle}{t(\|(A + tX)\xi\| + \|A\xi\|)} \\
&= \frac{\langle A\xi, X\xi \rangle + \langle X\xi, A\xi \rangle}{2\|A\|}.
\end{aligned}
$$

Hence the necessary condition implies that the above set on the right reduces to a singleton which gives the right directional derivative. We also have

$$
\begin{aligned}
\lim_{t \uparrow 0} \frac{\|(A + tX)\xi\| - \|A\xi\|}{t} &= -\lim_{t \downarrow 0} \frac{\|(A - tX)\xi\| - \|A\xi\|}{t} \\
&= -\frac{\langle A\xi, -X\xi \rangle + \langle -X\xi, A\xi \rangle}{2\|A\|} \\
&= \lim_{t \downarrow 0} \frac{\|(A + tX)\xi\| - \|A\xi\|}{t}.
\end{aligned}
$$

This proves Gateaux differentiability at A. The last assertion is clear from the above computation.   □

Example 2.2.2. Let $A = \begin{pmatrix} 1 & 0 \\ 0 & 0 \end{pmatrix} \in M_2$. Then the unit vectors in $\mathbb{C}^2$ where A achieves its norm are of the form $(\alpha, 0)$ with $|\alpha| = 1$. For any matrix $X = (x_{ij})$ in M_2, we have $\langle A(\alpha, 0)^T, X(\alpha, 0)^T \rangle = \overline{x_{12}} + \overline{x_{21}}$ which is independent of α, and the C*-norm is Gateaux differentiable at A with derivative

$$\partial \|A\|(X) = \mathrm{Re}\, \langle A(1,0)^T, X(1,0)^T \rangle = \mathrm{Re}\, x_{11}.$$

The matrix $B = \begin{pmatrix} 1 & 1 \\ 0 & 0 \end{pmatrix}$ achieves its norm at $(\sqrt{2}, 0)$ and $(\frac{\sqrt{2}}{2}, \frac{\sqrt{2}}{2})$; but

$$\left\langle B(\sqrt{2}, 0)^T, X(\sqrt{2}, 0)^T \right\rangle \neq \left\langle B\left(\frac{\sqrt{2}}{2}, \frac{\sqrt{2}}{2}\right)^T, X\left(\frac{\sqrt{2}}{2}, \frac{\sqrt{2}}{2}\right)^T \right\rangle$$

if X is the identity matrix, say. Hence the C*-norm is not Gateaux differentiable at B, however, we have the *right I-directional* derivative

$$\partial^+ \|B\|(I) = \lim_{t \downarrow 0} \frac{\|B + tI\| - \|B\|}{t}$$

$$= \lim_{t \downarrow 0} \frac{\sqrt{1 + t + t^2} + \sqrt{1 + 2t + 2t^2} - \sqrt{2}}{t} = \frac{\sqrt{2}}{2}.$$

On the other hand, the trace norm $\| \cdot \|_{tr}$ is not Gateaux differentiable at A since

$$\frac{\|A + tX\|_{tr} - \|A\|_{tr}}{t} = \frac{|t|}{t}$$

for $X = \begin{pmatrix} 0 & 0 \\ 0 & -1 \end{pmatrix}$, say.

Lemma 2.2.3. *Let $A \in M_n \setminus \{0\}$ with polar decomposition $A = u|A|$. If the trace norm $\| \cdot \|_{tr}$ on M_n is Gateaux differentiable at A, then the Gateaux derivative is given by*

$$\partial \|A\|_{tr}(X) = \mathrm{Re}\, \mathrm{Tr}(u^* X) \qquad (X \in M_n).$$

Proof. We only need to show that $\psi(X) = \mathrm{Re}\, Tr(u^* X)$ is a subdifferential. Indeed, we have $|A| = u^* A$ and

$$\psi(X - A) = \mathrm{Re}\, \mathrm{Tr}(u^* X) - \mathrm{Re}\, T_1(u^* A)$$
$$\leq \|u^*\| \|X\|_{tr} - \|A\|_{tr}$$
$$= \|X\|_{tr} - \|A\|_{tr}.$$

       □

Example 2.2.4. In Example 2.2.2 above, we have $u = A$ in the polar decomposition of A and $\operatorname{Re} Tr(u^*X) = 0$ for $X = \begin{pmatrix} 0 & 0 \\ 0 & -1 \end{pmatrix}$, while the *right* X-directional derivative is given by

$$\lim_{t \downarrow 0} \frac{\|A + tX\|_{tr} - \|A\|_{tr}}{t} = \lim_{t \downarrow 0} \frac{|t|}{t} = 1.$$

Due to the non-smoothness of the C*-norm and trace norm on M_n, we will consider the Lebesgue spaces $L^p(G, (M_n, \|\cdot\|_{hs}))$ with M_n equipped with the Hilbert-Schmidt norm when we need to make use of norm differentiability later. We compute below the Gateaux derivatives for $L^p(G, (M_n, \|\cdot\|_{hs}))$.

Since the function $u \in E \mapsto \|u\|^p$ is convex on any Banach space E, we have, for $0 < t < 1$ and $u, v \in E$,

$$\|u + tv\|^p \le (1 - t)\|u\|^p + t\|u + v\|^p$$

and

$$\|u\|^p \le \frac{t}{1+t}\|u - v\|^p + \frac{1}{1+t}\|u + tv\|^p$$

which gives

$$\|u\|^p - \|u - v\|^p \le \frac{1}{t}(\|u + tv\|^p - \|u\|^p) \le \|u + v\|^p - \|u\|^p. \qquad (2.4)$$

Proposition 2.2.5. *Let* $1 < p < \infty$. *The norm of* $L^p(G, (M_n, \|\cdot\|_{hs}))$ *is Gateaux differentiable at each non-zero f with Gateaux derivative*

$$\partial\|f\|_p(g) = \operatorname{Re}\|f\|_p^{1-p} \int_{\{x:f(x) \ne 0\}} \|f(x)\|_{hs}^{p-2} \operatorname{Tr}(f(x)^* g(x)) d\lambda(x)$$

for $g \in L^p(G, (M_n, \|\cdot\|_{hs}))$.

Proof. Given $A \in M_n \setminus \{0\}$, we have, by the chain rule,

$$\frac{d}{dt}\Big|_{t=0} \|A + tX\|_{hs}^p = p\|A\|_{hs}^{p-1} \frac{d}{dt}\Big|_{t=0} \|A + tX\|_{hs} = p\|A\|_{hs}^{p-1} \operatorname{Re} \operatorname{Tr}(A^*X)$$

for $X \in M_n$.

Fix a non-zero f in $L^p(G, (M_n, \|\cdot\|_{hs}))$. Given $p > 1$ and $g \in L^p(G, (M_n, \|\cdot\|_{hs}))$, we have

$$\frac{d}{dt}\Big|_{t=0} \|tg(x)\|_{hs}^p = 0.$$

By (2.4) and the dominated convergence theorem, we have

$$p\|f\|_p^{p-1} \frac{d}{dt}\bigg|_{t=0} \|f+tg\|_p = \frac{d}{dt}\bigg|_{t=0} \|f+tg\|_p^p$$

$$= \int_G \frac{d}{dt}\bigg|_{t=0} \|f(x)+tg(x)\|_{hs}^p d\lambda(x)$$

$$= \int_{\{x:f(x)\neq 0\}} \frac{d}{dt}\bigg|_{t=0} \|f(x)+tg(x)\|_{hs}^p d\lambda(x)$$

$$= \int_{\{x:f(x)\neq 0\}} p\|f(x)\|_{hs}^{p-2} \operatorname{Re} \operatorname{Tr}(f(x)^* g(x)) d\lambda(x)$$

which gives the formula for the Gateaux derivative at f. $\square$

Corollary 2.2.6. *For $1 < p < \infty$, the Lebesgue space $L^p(G, (M_n, \|\cdot\|_{hs}))$ is strictly convex, that is, the extreme points of its closed unit ball are exactly the functions of unit norm.*

Proof. This follows from the fact that a Banach space E is strictly convex if, and only if, the norm of its dual E^* is Gateaux differentiable on the unit sphere. $\square$

Chapter 3
Matrix Convolution Operators

In this Chapter, we study the basic structures of matrix convolution operators $T_\sigma : f \in L^p(G, M_n) \mapsto f * \sigma \in L^p(G, M_n)$. Noncommutativity of the matrix multiplication necessitates the introduction of the *left* convolution operator $L_\sigma : f \mapsto \sigma *_\ell f$ for a consistent duality theory. We first characterise these operators and show they are translation invariant operators satisfying some continuity condition. We also determine when these operators are weakly compact on L^1 and L^∞ spaces.

Spectral theory is developed in Section 3. We introduce the matrix-valued Fourier transform and, for abelian groups, the determinant of a matrix-valued measure. We describe the L^p spectrum of T_σ in Theorem 3.3.23, for an absolutely continuous matrix measure σ on an abelian group. For non-abelian groups, we develop a device to study the L^2 spectrum by identifying L_σ as an element in the tensor product $C_r^*(G) \otimes M_n$ of the reduced group C*-algebra $C_r^*(G)$ and M_n. We show that, for absolutely continuous symmetric σ, the L^2 spectrum of T_σ is the union of the spectrum of each element in $\widehat{\sigma}(\widehat{G}_r)$, where $\widehat{\sigma}$ is the Fourier transform of σ and $\widehat{G}_r$ is the reduced dual of G. This result is used to compute the spectrum of a homogeneous graph.

We study eigenspaces of T_σ in the latter part of the Chapter. The focus is on the eigenspace

$$H_\alpha(T_\sigma, L^p) = \{ f \in L^p(G, M_n) : T_\sigma(f) = \alpha f \}$$

with $\alpha = \|\sigma\|$, which is the space of matrix L^p harmonic functions on G. We show that $H_{\|\sigma\|}(T_\sigma, L^p)$ is the range of a contractive projection P on $L^p(G, M_n)$ for $1 < p < \infty$, extending an analogous result in [9, 13] for $p = \infty$. We discuss Liouville theorem and Poisson representation for harmonic functions and show that $\dim H_{\|\sigma\|}(T_\sigma, L^1) \le n^2$ if σ is positive and adapted. As an application, we use the results to show the existence of L^∞ non-constant harmonic functions on Riemannian symmetric spaces of non-compact type. Finally, we study the Jordan structures in $H_{\|\sigma\|}(T_\sigma, L^\infty)$ and in particular, determine when it is a Jordan subtriple of $L^\infty(G)$ in the scalar case.

C.-H. Chu, *Matrix Convolution Operators on Groups*. Lecture Notes in Mathematics 1956,   21
doi: 10.1007/978-3-540-69798-5, © Springer-Verlag Berlin Heidelberg 2008

3.1 Characterisation of Matrix Convolution Operators

We are now prepared to study the structures of matrix-valued convolution operators $f \mapsto f * \sigma$ on the Lebesgue spaces $L^p(G, M_n)$ of matrix functions on a locally compact group G, induced by a matrix-valued measure σ on G. Matrix convolutions of distributions in $\mathbb{R}^n$ have also been considered in [49].

We begin in this section by characterising matrix convolution operators. We show they are the translation invariant operators satisfying some continuity condition. On $L^1(G, M_n)$, they are exactly the translation invariant operators, and on $L^\infty(G, M_n)$ they are the weak* continuous translation invariant operators.

We note that M_n is equipped with the C*-norm throughout unless otherwise stated. A matrix $A \in M_n$ is *positive* if $\langle A\xi, \xi \rangle \geq 0$ for all vectors $\xi \in \mathbb{C}^n$. Let M_n^+ denote the cone of positive matrices in M_n.

We first introduce the notion of a matrix-valued measure. By an M_n-*valued measure* μ on a locally compact space G, we mean a (norm) countably additive function $\mu : \mathcal{B} \to M_n$ where $\mathcal{B}$ is the σ-algebra of Borel sets in G. Since the trace-norm $\| \cdot \|_{tr}$ is equivalent to the C*-algebra norm on M_n and $M_n^* = (M_n, \| \cdot \|_{tr})$, we can regard an M_n^*-valued measure on G as an M_n-valued measure on G, and *vice versa*. We can denote an M_n-valued measure μ on G by an $n \times n$ matrix $\mu = (\mu_{ij})$ of complex-valued measures μ_{ij} on G. The variation $|\mu|$ of μ is a positive real finite measure on G defined by

$$|\mu|(E) = \sup_{\mathcal{P}} \left\{ \sum_{E_i \in \mathcal{P}} \|\mu(E_i)\| \right\} \qquad (E \in \mathcal{B})$$

with the supremum taken over all partitions $\mathcal{P}$ of E into a finite number of pairwise disjoint Borel sets. We define the norm of μ to be $\|\mu\| = |\mu|(G)$. As shown in [9, p. 21], μ has a polar representation $\mu = \omega \cdot |\mu|$ where $\omega : G \to M_n$ is a Bochner integrable function with $\|\omega(\cdot)\| = 1$. Likewise, if μ is an M_n^*-valued measure, we define its norm by $\|\mu\|_{tr} = |\mu|_{tr}(G) = \sup_{\mathcal{P}} \{ \sum_{E_i \in \mathcal{P}} \|\mu(E_i)\|_{tr} \}$.

A function $f = (f_{ij}) : G \longrightarrow M_n$ is said to be μ-*integrable* if each f_{ij} is a Borel function and the integrals $\int_G f_{ij} d\mu_{k\ell}$ exist in which case, we define, for any $E \in \mathcal{B}$, the *integral* $\int_E f d\mu$ to be an $n \times n$ matrix with ij-th entry

$$\sum_k \int_E f_{ik} d\mu_{kj}.$$

We have

$$\left\| \int_E f d\sigma \right\| = \left\| \int_E f(x)\omega(x) d|\sigma|(x) \right\| \leq \int_E \|f(x)\| d|\sigma|(x) \qquad (3.1)$$

since $\|\omega(\cdot)\| = 1$. If we regard an M_n-valued measure μ as an M_n^*-valued measure, then we can also regard an M_n-valued μ-integrable function f on G as an M_n^*-valued μ-integrable function, with $\int_E f d\mu \in M_n^*$, and *vice versa*.

Let $M(G, M_n^*)$ be the space of all M_n^*-valued measures on G, equipped with the total variation norm $\| \cdot \|_{tr}$. It is linearly isomorphic to the space $M(G, M_n)$ of M_n-valued measures on G, equipped with the total variation norm $\| \cdot \|$. Let $C_0(G, M_n)$ be the Banach space of continuous M_n-valued functions on G vanishing at infinity, equipped with the supremum norm. Let $C_c(G, M_n)$ be the subspace of $C_0(G, M_n)$ consisting of continuous M_n-valued functions with compact support. We denote by $C_b(G, M_n)$ the Banach space of bounded continuous M_n-valued functions on G.

It has been shown in [9, Lemma 5] that $M(G, M_n^*)$ is linearly isometric order-isomorphic to the dual of $C_0(G, M_n)$, where a measure $\mu \in M(G, M_n^*)$ and a function $f \in C_0(G, M_n)$ are *positive* if $\mu(E)$ and $f(x)$ are positive matrices for all $E \in \mathcal{B}$ and $x \in G$. The above duality is given by

$$\langle , \rangle : C_0(G, M_n) \times M(G, M_n^*) \to \mathbb{C}$$

$$\langle f, \mu \rangle = \mathrm{Tr}\left(\int_G f d\mu \right) = \sum_{i,k} \int_G f_{ik} d\mu_{ki}$$

where $f = (f_{ij}) \in C_0(G, M_n)$ and $\mu = (\mu_{ij}) \in M(G, M_n^*)$ (cf. [9, Lemma 5]). Further, as shown in [14, Proposition 2.4], $(M(G, M_n^*), \| \cdot \|_{tr})$ is a Banach algebra in the convolution product $\mu * \nu$ given by

$$\langle f, \mu * \nu \rangle = \mathrm{Tr}\left(\int_G \int_G f(xy) d\mu(x) d\nu(y) \right) \qquad (f \in C_0(G, M_n)).$$

Likewise $(M(G, M_n), \| \cdot \|)$ is a Banach algebra in the convolution product and is algebraically isomorphic to $M(G, M_n^*)$.

Given $a \in G$, we denote by $\delta_a \in M(G, M_n)$ the unit mass at a, having values $\{0, I\} \subset M_n$. A measure $\sigma \in M(G, M_n)$ is called *adapted* if its variation $|\sigma|$ is an adapted measure on G, that is, $\mathrm{supp}\,|\sigma|$ generates a dense subgroup of G.

Since the matrix product is non-commutative, given a matrix valued measure $\sigma = (\sigma_{ij})$ and a matrix Borel function $f = (f_{ij})$ on G, besides the matrix-valued integral $\int_G f d\sigma$ defined above, we need to introduce the *transposed* integral

$$\int_G d\sigma(x) f(x)$$

which is defined to have the ij-entry

$$\left(\int_G d\sigma(x) f(x) \right)_{ij} = \sum_k \int_G f_{kj}(x) d\sigma_{ik}(x).$$

We also have

$$\left\| \int_G d\sigma(x) f(x) \right\| \leq \int_G \|f(x)\| d|\sigma|(x). \tag{3.2}$$

The matrix-valued convolution $f * \sigma$, if exists at $x \in G$, is defined by

$$f * \sigma(x) = \int_G f(xy^{-1}) d\sigma(y).$$

We note that the same definition of matrix convolution of a matrix-valued distribution and measure on $\mathbb{R}^n$ has been given in [49, p.279].

We introduce the *left* convolution $\sigma *_\ell f$ to be the following integral if it exists:

$$\sigma *_\ell f(x) = \int_G d\sigma(y) f(xy^{-1}) \qquad (x \in G).$$

The subscript ℓ is used to avoid confusion with the convolution $\sigma * f$ in the scalar case:

$$\sigma * f(x) = \int_G f(y^{-1}x) \triangle_G(y^{-1}) d\sigma(y).$$

Given $\sigma \in M(G, M_n)$, we define the measure $\widetilde{\sigma} \in M(G, M_n)$ by $d\widetilde{\sigma}(x) = d\sigma(x^{-1})$, as in the scalar case.

For complex measures μ and σ, we have

$$\widetilde{\mu * \sigma} = \widetilde{\sigma} * \widetilde{\mu}$$

for $\mu, \sigma \in M(G)$. This formula need not hold for matrix-valued measures $\mu, \sigma \in M(G, M_n)$, instead, we have

$$\widetilde{\mu * \sigma} = \widetilde{\mu} *_\ell \widetilde{\sigma} \tag{3.3}$$

where the *transposed* convolution $\widetilde{\mu} *_\ell \widetilde{\sigma}$ is defined by

$$\widetilde{\mu} *_\ell \widetilde{\sigma}(f) = \mathrm{Tr} \left(\int_G \int_G f(xy) d\widetilde{\mu}(y) d\widetilde{\sigma}(x) \right) \qquad (f \in C_0(G, M_n)).$$

If $\sigma = g \cdot \lambda$ for some $g \in L^1(G, M_n)$, then we have

$$\widetilde{\mu} *_\ell \widetilde{\sigma} = (\widetilde{\mu} *_\ell g) \cdot \lambda.$$

This is one of the reasons for introducing transposed integrals.

An M_n-valued measure σ is called *absolutely continuous* (w.r.t. λ) if its total variation $|\sigma|$ is absolutely continuous with respect to the Haar measure λ. This is equivalent to the existence of a function $h \in L^1(G, M_n)$ such that $\sigma = h \cdot \lambda$. Indeed, given the latter and given $\lambda(E) = 0$ for some Borel set $E \subset G$, we have

$$|\sigma|(E) = \int_E \|\omega(\cdot)\|^2 d|\sigma| = \int_E \|\omega(\cdot)^* \omega(\cdot)\| d|\sigma|$$

$$\leq \int_E \mathrm{Tr}(\omega(\cdot)^* \omega(\cdot)) d|\sigma| = \mathrm{Tr} \int_E \omega^* d\sigma$$

$$= \mathrm{Tr} \int_E \omega^* h d\lambda = 0.$$

Conversely, if $|\sigma| = k \cdot \lambda$ for some $k \in L^1(G)$, then $k\omega \in L^1(G,M_n)$ and $\sigma = k\omega \cdot \lambda$.

Let $1/p + 1/q = 1$. Since M_n has the Radon-Nikodym property, the dual $L^p(G,M_n)^*$ identifies with the space $L^q(G,M_n^*)$ for $1 \leq p < \infty$, with the duality

$$\langle \cdot, \cdot \rangle : L^p(G,M_n) \times L^q(G,M_n^*) \longrightarrow \mathbb{C}$$

given by

$$\langle f, h \rangle = \mathrm{Tr}\left(\int_G f(x)h(x)d\lambda(x)\right)$$

(cf. [22, p.98] and [34]). Likewise $L^p(G,M_n^*)^* = L^q(G,M_n)$.

For $f \in L^p(G,M_n)$ and $h \in L^q(G,M_n^*)$, we have

$$\langle f * \sigma, h \rangle = \mathrm{Tr}\left(\int_G \int_G h(xy)f(x)d\sigma(y)d\lambda(x)\right) = \langle f, \widetilde{\sigma} *_\ell h \rangle \qquad (3.4)$$

which is identical to (2.2) in the scalar case. The above duality is another *raison d'être* for the transposed integral $\sigma *_\ell f$.

As before, given a function $h : G \longrightarrow M_n$, we define $\widetilde{h}(x) = h(x^{-1})$.

Lemma 3.1.1. *Let $f \in L^p(G,M_n)$ and $\psi \in L^q(G,M_n)$ where $1 \leq p,q \leq \infty$. Then $\widetilde{\psi} * f : G \longrightarrow M_n$ is a bounded and left uniformly continuous function.*

Proof. Since $\widetilde{\psi} * f$ has the ij-entry

$$(\widetilde{\psi} * f)_{ij} = \sum_k \widetilde{\psi}_{ik} * f_{kj},$$

the result follows from entry-wise application of the scalar result in [10, Lemma 3.2]. □

We have

$$\langle f, h \rangle = \mathrm{Tr}(\widetilde{h} * f)(e) \qquad (f \in L^p(G,M_n), \ h \in L^q(G,M_n))$$

and hence the following consequence.

Lemma 3.1.2. *If $f \in L^p(G,M_n)$ and $\mathrm{Tr}(\widetilde{h} * f)(e) = 0$ for all $h \in L^q(G,M_n)$, then $f = 0$.*

Given $f \in L^1(G,M_n)$, the measure $f \cdot \lambda \in M(G,M_n)$ has total variation

$$\|f \cdot \lambda\| = |f \cdot \lambda|(G) = \int_G \|f(x)\|d\lambda(x) = \|f\|_1$$

by [22, p.46] and we can therefore, as in the scalar case, identify $L^1(G,M_n)$ as a closed subspace of $M(G,M_n)$, consisting of absolutely continuous M_n-valued measures on G. Moreover, $L^1(G,M_n)$ is a right ideal of the Banach algebra $M(G,M_n)$ since

$$(f \cdot \lambda) * \mu = (f * \mu) \cdot \lambda$$

for $f \in L^1(G,M_n)$ and $\mu \in M(G,M_n)$. Likewise, we identify $L^1(G,M_n^*)$ as an ideal in the Banach algebra $M(G,M_n^*)$.

We now define matrix convolution operators. For $1 \leq p \leq \infty$ and $\sigma \in M(G,M_n)$, we define $T_\sigma : L^p(G,M_n) \longrightarrow L^p(G,M_n)$ by

$$T_\sigma(f) = f * \sigma \qquad (f \in L^p(G,M_n)).$$

We also define the *left* convolution operator $L_\sigma : L^p(G,M_n) \longrightarrow L^p(G,M_n)$ by

$$L_\sigma(f) = \sigma *_\ell f \qquad (f \in L^p(G,M_n)).$$

To avoid triviality, σ is always non-zero for T_σ and L_σ. The operators T_σ and L_σ are well-defined since, as in the scalar case [10, Lemma 2.1], one can show that $\|f * \sigma\|_p \leq \|f\|_p \|\sigma\|$ and $\|\sigma *_\ell f\|_p \leq \|f\|_p \|\sigma\|$, using (3.1) and (3.2).

In contrast to Lemma 2.1.2 where a convolution operator T_σ on $L^\infty(G)$ has norm $\|T_\sigma\|_\infty = \|\sigma\|$ for a complex measure σ, it is possible to have $\|T_\sigma\|_\infty < \|\sigma\|$ if σ is *matrix-valued* for T_σ defined on $L^\infty(G,M_n)$.

Example 3.1.3. Let $G = \{a,e\}$ and define a positive M_2-valued measure σ on G by

$$\sigma\{a\} = \begin{pmatrix} 1 & 0 \\ 0 & 0 \end{pmatrix} \quad \text{and} \quad \sigma\{e\} = \begin{pmatrix} 0 & 0 \\ 0 & 2 \end{pmatrix}.$$

Then $|\sigma|(G) = \|\sigma\{a\}\| + \|\sigma\{e\}\| = 3$.

Let $f \in L^\infty(G,M_2)$ with $f(a) = (a_{ij})$ and $f(e) = (b_{ij})$. If $\|f\|_\infty \leq 1$, then $|a_{12}|^2 + |a_{22}|^2 \leq 1$ and $|b_{11}|^2 + |b_{21}|^2 \leq 1$. Hence

$$\|f * \sigma(a)\| = \left\| \begin{pmatrix} b_{11} & 2a_{12} \\ b_{21} & 2a_{22} \end{pmatrix} \right\|$$

$$\leq (|b_{11}|^2 + |b_{21}|^2 + 4|a_{12}|^2 + 4|a_{22}|^2)^{\frac{1}{2}} \leq \sqrt{5}.$$

Likewise, we have

$$\|f * \sigma(e)\| = \left\| \begin{pmatrix} a_{11} & 2b_{12} \\ a_{21} & 2b_{22} \end{pmatrix} \right\| \leq \sqrt{5}.$$

Therefore we have $\|f * \sigma\|_\infty \leq \sqrt{5} < 3 = \|\sigma\|$ and $\|T_\sigma\|_\infty < \|\sigma\|$. This difference from the scalar case is due to the presence of the trace Tr in the norm of σ:

$$\|\sigma\| = \sup \left\{ \left| \mathrm{Tr} \int_G f d\sigma \right| : f \in C_0(G,M_n) \text{ and } \|f\| \leq 1 \right\}.$$

Although we have $\| \int_G f d\sigma \| \leq \|T_\sigma\|_\infty$, the value $|\mathrm{Tr} \int_G f d\sigma| = |a_{11}| + 2|b_{22}|$ could exceed $\|T_\sigma\|_\infty$.

Lemma 3.1.4. *Let $\sigma \in M(G,M_n)$. Then for $p < \infty$, the dual map of T_σ : $L^p(G,M_n) \longrightarrow L^p(G,M_n)$ is the convolution operator $L_{\tilde\sigma} : L^q(G,M_n^*) \longrightarrow L^q(G,M_n^*)$.*

The weak* continuous operator $T_\sigma : L^\infty(G, M_n) \longrightarrow L^\infty(G, M_n)$ has predual $L_{\widetilde{\sigma}} : L^1(G, M_n^*) \longrightarrow L^1(G, M_n^*)$.

Proof. This follows from the duality in (3.4). ☐

As noted before, $L^p(G, M_n)$ is linearly isomorphic to $L^p(G, M_n^*)$ and the latter is identified with $L^p(G, (M_n, \| \cdot \|_{tr}))$, equipped with the norm

$$\|f\|_p = \left(\int_G \|f(x)\|_{tr}^p d\lambda(x) \right)^{\frac{1}{p}}$$

for $1 \leq p < \infty$, and likewise for $L^\infty(G, M_n^*)$. Hence a continuous linear map $T :$ $L^p(G, M_n) \longrightarrow L^p(G, M_n)$ can be regarded as one on $L^p(G, M_n^*)$, and *vice versa*, although the norm $\|T\|_{L^p(G,M_n)}$ differs from $\|T\|_{L^p(G,M_n^*)}$ in general. Nevertheless, $T_{L^p(G,M_n)}$ and $T_{L^p(G,M_n^*)}$ have the same spectrum.

We regard $L^p(G, M_n)$ as a left M_n-module by defining

$$(Af)(x) = Af(x) \qquad (x \in G)$$

for $A \in M_n$ and $f \in L^p(G, M_n)$. In this case, every convolution operator T_σ is M_n-*linear*, that is, T_σ is complex linear as well as

$$T_\sigma(Af) = AT_\sigma(f) \qquad (A \in M_n, f \in L^p(G, M_n)).$$

Also T_σ is invariant under left translations:

$$\ell_x T_\sigma = T_\sigma \ell_x \qquad (x \in G).$$

We characterise below T_σ among left-translation invariant M_n-linear operators on $L^p(G, M_n)$.

We note that, since we adopt the *right* Haar measure on G, the right translation $r_x :$ $L^p(G, M_n) \longrightarrow L^p(G, M_n)$ is an isometry and the dual $r_x^* : L^q(G, M_n^*) \longrightarrow L^q(G, M_n^*)$ satisfies $r_x^* = r_{x^{-1}}$ whereas the left translation ℓ_x has dual $\ell_x^* = \triangle_G(x)\ell_{x^{-1}}$. For $1 \leq p < \infty$, we have $\|\ell_x(f)\|_p = \triangle_G(x)\|f\|_p$.

Lemma 3.1.5. *Let $B \in M_n$. If $\mathrm{Tr}(BA) = 0$ for every positive invertible matrix $A \in M_n$, then $B = 0$.*

Proof. We have $\mathrm{Tr}(B) = 0$. For every positive matrix A, the matrix $A + \frac{1}{n}I$ is invertible and we have $\mathrm{Tr}(BA) = \mathrm{Tr}(B(A + \frac{1}{n}I)) = 0$. Hence $B = 0$. ☐

Evidently $C_0(G, M_n)$ is also a left M_n-module.

Lemma 3.1.6. *Let $\psi : C_0(G, M_n) \longrightarrow M_n$ be a continuous M_n-linear map. Then there is a unique $\sigma \in M(G, M_n^*)$ such that*

$$\psi(f) = \int_G f d\sigma \qquad (f \in C_0(G, M_n)).$$

Proof. Let A be a positive invertible matrix in M_n. Define a continuous linear functional $\varphi \in C_0(G,M_n)^*$ by

$$\varphi(f) = \text{Tr}(A\psi(f)) \qquad (f \in C_0(G,M_n)).$$

By [9, Lemma 5], there is a unique measure $\mu_A \in M(G,M_n^*)$ such that

$$Tr(A\psi(f)) = Tr\left(\int_G f d\mu_A\right) \qquad (f \in C_0(G,M_n))$$

which implies

$$Tr(\psi(f)) = \varphi(A^{-1}f) = Tr\left(\int_G A^{-1} f d\mu_A\right) = Tr\left(\int_G f d\mu_A A^{-1}\right) \qquad (3.5)$$

for $f \in C_0(G,M_n)$. Define $\sigma \in M(G,M_n^*)$ by $\sigma(E) = \mu_A(E)A^{-1}$ for each Borel set $E \subset G$. By (3.5), we have

$$\text{Tr}\left(\int_G f d\mu_A A^{-1}\right) = \text{Tr}\left(\int_G f d\mu_B B^{-1}\right) \qquad (f \in C_0(G,M_n))$$

for every positive invertible matrix $B \in M_n$. By the isomorphism between $C_0(G,M_n)^*$ and $M(G,M_n^*)$, we have $\sigma = \mu_B B^{-1}$ for every positive invertible $B \in M_n$ from (3.5), and also

$$\text{Tr}(\psi(f)B) = \text{Tr}\left(\int_G f d\sigma B\right) \qquad (f \in C_0(G,M_n)).$$

Therefore we have

$$\psi(f) = \int_G f d\sigma \qquad (f \in C_0(G,M_n))$$

by Lemma 3.1.5. The uniqueness of σ is clear. $\qquad\qquad\square$

Proposition 3.1.7. *Let* $T : L^\infty(G,M_n) \longrightarrow L^\infty(G,M_n)$ *be a bounded M_n-linear map satisfying*

$$\ell_x T = T\ell_x \qquad (x \in G).$$

Then there is a unique measure $\sigma \in M(G,M_n)$ *such that*

$$Tf = f * \sigma \quad \text{for} \quad f \in C_0(G,M_n).$$

Proof. Let (u_β) be a bounded approximate identity in $L^1(G)$. Then

$$v_\beta = \begin{pmatrix} u_\beta & & \\ & \ddots & \\ & & u_\beta \end{pmatrix}$$

is a bounded approximate identity in $L^1(G, M_n)$. Let $\mu_\beta = \nu_\beta \cdot \lambda$. Since ν_β is diagonal, one verifies readily that $\mu_\beta *_\ell h = h * \nu_\beta$ for $h \in L^1(G, M_n)$. The measure $\widetilde{\mu_\beta} = \widetilde{\nu_\beta \triangle_G} \cdot \lambda$ is absolutely continuous, where $\widetilde{\nu_\beta \triangle_G} \in L^1(G)$.

Let $f \in C_0(G, M_n)$. Then we have $Tf * \widetilde{\mu_\beta} \in C_b(G, M_n)$ by Lemma 3.1.1. By (3.4), we have

$$\langle Tf * \widetilde{\mu_\beta}, h \rangle = \langle Tf, \mu_\beta *_\ell h \rangle = \langle Tf, h * \nu_\beta \rangle \qquad (h \in L^1(G, M_n))$$

which implies that the net $(Tf * \widetilde{\mu_\beta})$ weak*-converges to Tf in $L^\infty(G, M_n)$.

Define an M_n-linear map $\psi_\beta : C_0(G, M_n) \longrightarrow M_n$ by

$$\psi_\beta(f) = (Tf * \widetilde{\mu_\beta})(e) \qquad (f \in C_0(G, M_n)).$$

By Lemma 3.5, there is a unique measure $\sigma_\beta \in M(G, M_n^*)$, which can be regarded as an M_n-valued measure, such that

$$\psi_\beta(f) = \int_G f d\sigma_\beta \quad \text{for} \quad f \in C_0(G, M_n)$$

and for $x \in G$, we have

$$Tf * \widetilde{\mu_\beta}(x) = \ell_{x^{-1}}(Tf * \widetilde{\mu_\beta})(e) = (\ell_{x^{-1}} Tf * \widetilde{\mu_\beta})(e)$$
$$= (T(\ell_{x^{-1}} f) * \widetilde{\mu_\beta})(e) = \psi_\beta(\ell_{x^{-1}} f) = \int_G \ell_{x^{-1}} f d\sigma_\beta = f * \widetilde{\sigma}_\beta(x).$$

The net (σ_β) is norm bounded in $M(G, M_n)$ since

$$\|\sigma_\beta\| = \sup \left\{ \left| \text{Tr} \int_G f d\sigma_\beta \right| : f \in C_0(G, M_n) \text{ and } \|f\| \leq 1 \right\} \leq \|\text{Tr}\| \|T\| \|\nu_\beta\|_1.$$

By weak* compactness, (σ_β) has a subnet (σ_γ) weak*-converging to some $\sigma \in M(G, M_n)$.

Given $f \in C_c(G, M_n)$, the net $(f * \widetilde{\sigma}_\gamma)$ weak*-converges to $f * \widetilde{\sigma}$ in $L^\infty(G, M_n)$. Indeed, for each $h \in C_c(G, M_n) \subset L^1(G, M_n)$, we have

$$\langle h, f * \widetilde{\sigma}_\gamma \rangle = \text{Tr} \int_G \widetilde{h} * f d\sigma_\gamma \longrightarrow \langle h, f * \widetilde{\sigma} \rangle$$

where $\widetilde{h} * f \in C_c(G, M_n)$. Note that the net $(f * \widetilde{\sigma}_\gamma)$ is norm bounded in $L^\infty(G, M_n)$ since $\|f * \widetilde{\sigma}_\gamma\| \leq \|f\|_\infty \|\widetilde{\sigma}_\gamma\|$. By density of $C_c(G, M_n)$ in $L^1(G, M_n)$, we deduce that

$$\langle k, f * \widetilde{\sigma}_\gamma \rangle \longrightarrow \langle k, f * \widetilde{\sigma} \rangle$$

for each $k \in L^1(G, M_n)$.

Hence for each $f \in C_c(G, M_n)$, we have

$$Tf = \text{w*-}\lim_\gamma Tf * \widetilde{\mu_\gamma} = \text{w*-}\lim_\gamma f * \widetilde{\sigma}_\gamma = f * \widetilde{\sigma}$$

which implies $Tf = f * \widetilde{\sigma}$ for all $f \in C_0(G, M_n)$. Finally, the uniqueness of $\widetilde{\sigma}$ is evident. $\square$

Remark 3.1.8. In the conclusion of the above result, we cannot expect $T = T_\sigma$ on $L^\infty(G, M_n)$ in general. In fact, even in the case of a translation invariant operator T on $\ell^\infty(\mathbb{Z})$, $T \neq T_\sigma$ can occur on $C_b(\mathbb{Z})$ [44, p.78].

We have the following characterization of convolution operators on $L^\infty(G, M_n)$.

Proposition 3.1.9. *Let* $T : L^\infty(G, M_n) \longrightarrow L^\infty(G, M_n)$ *be a bounded M_n-linear map. The following conditions are equivalent.*

(i) $T = T_\sigma$ *for some* $\sigma \in M(G, M_n)$.
(ii) T *is weak* continuous and* $\ell_x T = T \ell_x$ *for all* $x \in G$.

Proof. We need only prove (ii) $\Longrightarrow$ (i). By weak* continuity, T has a predual $T_* :$ $L^1(G, M_n^*) \longrightarrow L^1(G, M_n^*)$. By Proposition 3.1.7, there is a measure $\sigma \in M(G, M_n)$ such that
$$Tg = g * \sigma \quad \text{for} \quad g \in C_0(G, M_n).$$
We use the duality $C_0(G, M_n)^* = M(G, M_n^*)$. Let $f \in L^1(G, M_n^*) \subset M(G, M_n^*)$. Identify both $T_* f$ and $\widetilde{\sigma} *_\ell f$ as absolutely continuous measures in $M(G, M_n^*)$. For $g \in C_0(G, M_n)$, we have, by (3.4),
$$\langle g, T_* f \rangle = \langle Tg, f \rangle = \langle g * \sigma, f \rangle = \langle g, \widetilde{\sigma} *_\ell f \rangle.$$
Hence $T_* f = \widetilde{\sigma} *_\ell f$. As f was arbitrary, this gives $T_* = L_{\widetilde{\sigma}}$ and $T = T_\sigma$. $\square$

Next we characterize convolution operators on $L^p(G, M_n)$ for $p < \infty$.

Proposition 3.1.10. *Let* $1 \leq p < \infty$ *and let* $T : L^p(G, M_n) \longrightarrow L^p(G, M_n)$ *be a bounded M_n-linear map. The following conditions are equivalent.*

(i) $T = T_\sigma$ *for some* $\sigma \in M(G, M_n)$.
(ii) $\ell_x T = T \ell_x$ *for all* $x \in G$ *and T maps $C_c(G, M_n)$ into $C_b(G, M_n)$ continuously in the supremum norm.*

Proof. We show (ii) $\Longrightarrow$ (i). Define an M_n-linear map $\psi : C_c(G, M_n) \longrightarrow M_n$ by
$$\psi(f) = Tf(e) \qquad (f \in C_c(G, M_n)).$$
Then ψ is continuous by condition (ii). Hence, as before, there is a measure $\sigma \in M(G, M_n)$ such that
$$\psi(f) = \int_G f d\sigma$$
for $f \in C_c(G, M_n)$ and we have
$$Tf(x) = \ell_{x^{-1}} Tf(e) = T(\ell_{x^{-1}} f)(e) = \psi(\ell_{x^{-1}} f) = \int_G \ell_{x^{-1}} f d\sigma = f * \widetilde{\sigma}(x).$$

For any $h \in L^p(G,M_n)$, there is a sequence (f_n) in $C_c(G,M_n)$ converging to h and therefore

$$Th = \lim_{n\to\infty} Tf_n = \lim_{n\to\infty} f_n * \tilde{\sigma} = h * \tilde{\sigma}.$$

□

We strengthen the above result for $p = 1$ in the next corollary.

Corollary 3.1.11. *Let* $T : L^1(G,M_n) \longrightarrow L^1(G,M_n)$ *be a bounded* M_n*-linear map. The following conditions are equivalent.*

(i) $T = T_\sigma$ *for some* $\sigma \in M(G,M_n)$.
(ii) $\ell_x T = T\ell_x$ *for all* $x \in G$.

Proof. For (ii) $\Longrightarrow$ (i), it suffices to show that T maps $C_c(G,M_n)$ into $C_b(G,M_n)$ continuously in the supremum norm. Since the dual $T^* : L^\infty(G,M_n^*) \longrightarrow L^\infty(G,M_n^*)$ is weak* continuous and commutes with left translations, we have $T^* = T_\mu$ for some $\mu \in M(G,M_n^*)$, by Proposition 3.1.9. It follows that $T = L_{\tilde{\mu}}$ which maps $C_c(G,M_n)$ into $C_b(G,M_n)$ continuously in the supremum norm. □

Corollary 3.1.12. *Let* G *be a compact group and let* $T : L^\infty(G,M_n) \longrightarrow L^\infty(G,M_n)$ *be a bounded* M_n*-linear map. The following conditions are equivalent.*

(i) $T = T_\sigma$ *for some* $\sigma \in M(G,M_n)$.
(ii) $\ell_x T = T\ell_x$ *for all* $x \in G$.

Proof. For (ii) $\Longrightarrow$ (i), we prove that T can be extended to a left-translation invariant operator on $L^1(G,M_n)$ and hence Corollary 3.1.11 applies.

Given that G is compact, we have $L^\infty(G,M_n) \subset L^1(G,M_n)$. By Proposition 3.1.7, there exists $\mu \in M(G,M_n)$ such that $Tf = f * \mu$ for all $f \in C(G,M_n)$ and hence

$$\|Tf\|_1 = \|f * \mu\|_1 \leq \|\mu\|\|f\|_1.$$

Since $C(G,M_n)$ is $\|\cdot\|_1$-dense in $L^1(G,M_n)$, we can extend T to an M_n-linear operator on $L^1(G,M_n)$ commuting with left translations. □

3.2 Weak Compactness of Convolution Operators

For $1 < p < \infty$, the operator $T_\sigma : L^p(G,M_n) \longrightarrow L^p(G,M_n)$ is weakly compact. In this section, we determine weak compactness conditions for T_σ on $L^1(G,M_n)$ and $L^\infty(G,M_n)$.

Theorem 3.2.1. *Let* $\sigma \in M(G,M_n)$ *be a positive measure such that* $|\sigma|^2 * |\tilde{\sigma}|^2$ *is adapted on* G. *The following conditions are equivalent.*

(i) $T_\sigma : L^1(G,M_n) \longrightarrow L^1(G,M_n)$ *is weakly compact.*
(ii) $T_\sigma : L^\infty(G,M_n) \longrightarrow L^\infty(G,M_n)$ *is weakly compact.*

(iii) $L_\sigma : L^1(G,M_n) \longrightarrow L^1(G,M_n)$ is weakly compact.
(iv) $L_\sigma : L^\infty(G,M_n) \longrightarrow L^\infty(G,M_n)$ is weakly compact.
 (v) $T_\sigma : L^1(G,M_n) \longrightarrow L^1(G,M_n)$ is compact.
(vi) $T_\sigma : L^\infty(G,M_n) \longrightarrow L^\infty(G,M_n)$ is compact.
(vii) $T_\sigma : L^p(G,M_n) \longrightarrow L^p(G,M_n)$ is compact for all $p \in [1,\infty]$.
(viiii) G is compact and σ is absolutely continuous.

Proof. We first prove the equivalence of (i) and (viii). The equivalence (ii) $\Longleftrightarrow$ (viii) can be proved analogously.

Let $\sigma = (\sigma_{ij})$ and let $T_\sigma : L^1(G,M_n) \longrightarrow L^1(G,M_n)$ be weakly compact. We define the coordinate projection $P_{ij} : L^1(G,M_n) \longrightarrow L^1(G)$ by

$$P_{ij}(f_{ij}) = f_{ij}$$

for $(f_{ij}) \in L^1(G,M_n)$. Then P_{ij} is a contraction since

$$\int_G |f_{ij}(x)|d\lambda(x) \le \int_G \|f(x)\|d\lambda(x).$$

Let $D : L^1(G) \longrightarrow L^1(G,M_n)$ be the natural embedding

$$D(f) = \begin{pmatrix} f & & \\ & \ddots & \\ & & f \end{pmatrix}.$$

Then each scalar convolution operator $T_{\sigma_{ij}} : L^1(G) \longrightarrow L^1(G)$ is the composite $T_{\sigma_{ij}} = P_{ij}T_\sigma D$ and hence weakly compact.

Let $\mu \in M(G)$ be the positive measure

$$\mu = \mathrm{Tr} \circ \sigma = \sigma_{11} + \cdots + \sigma_{nn}.$$

We show that supp μ contains (and hence equals) supp $|\sigma|$. Let $x \in$ supp $|\sigma|$ and let V be an open subset of G containing x. We need to show $\mu(V) > 0$. Suppose $\mu(V) = 0$. Given a partition $V = \bigcup_k E_k$, the positivity of the matrices $\sigma(E_k)$ implies

$$\sum_k \|\sigma(E_k)\| \le \sum_k \mathrm{Tr}\sigma(E_k) \le \sum_k \mu(E_k) = \mu(V) = 0.$$

It follows that $|\sigma|(V) = 0$ which is a contradiction. Hence $|\mu|(V) > 0$ and this proves $x \in$ supp μ.

Likewise we have supp $\widetilde{\mu} \supset$ supp $|\sigma|$. It follows that

$$\mathrm{supp}\,\mu^2 * \widetilde{\mu}^2 = \overline{(\mathrm{supp}\,\mu)^2(\mathrm{supp}\,\widetilde{\mu})^2} \supset \overline{(\mathrm{supp}\,|\sigma|)^2(\mathrm{supp}\,|\widetilde{\sigma}|)^2)} = \mathrm{supp}\,|\sigma|^2 * |\widetilde{\sigma}|^2.$$

Hence $\mu^2 * \widetilde{\mu}^2$ is a positive adapted measure on G. The convolution operator

$$T_\mu = \sum_k T_{\sigma_{kk}} : L^1(G) \longrightarrow L^1(G)$$

is weakly compact. By Proposition 2.1.6, G is compact. By Remark 2.1.7, each complex measure σ_{ij} is absolutely continuous, say, $\sigma_{ij} = h_{ij} \cdot \lambda$ for some $h_{ij} \in L^1(G)$. It follows that $\sigma = (h_{ij}) \cdot \lambda$ is absolutely continuous.

Conversely, if G is compact and $\sigma = \omega \cdot |\sigma|$ is absolutely continuous. Then each $\sigma_{ij} = \omega_{ij} \cdot |\sigma|$ is absolutely continuous and by Proposition 2.1.6 again, the scalar convolution operator $T_{\sigma_{ij}} : L^1(G) \longrightarrow L^1(G)$ is weakly compact. Given $f = (f_{ij}) \in L^1(G, M_n)$, we have

$$(T_\sigma f)_{ij} = \sum_k f_{ik} * \sigma_{kj} = \sum_k T_{\sigma_{kj}}(f_{ik}).$$

If $\|f\|_1 \leq 1$, then $\int_G |f_{ij}(x)| d\lambda(x) \leq \int_G \|f(x)\| d\lambda(x) \leq 1$. Therefore entrywise computation implies that T_σ maps the closed unit ball of $L^1(G, M_n)$ onto a relatively weakly compact set, that is, T_σ is weakly compact.

The equivalence of (iii), (iv) and (viii) follows from the fact that, by Lemma 3.1.4, $L_\sigma : L^1(G, M_n) \longrightarrow L^1(G, M_n)$ has dual $T_{\widetilde\sigma} : L^\infty(G, M_n^*) \longrightarrow L^\infty(G, M_n^*)$ and $L_\sigma : L^\infty(G, M_n) \longrightarrow L^\infty(G, M_n)$ has predual $T_{\widetilde\sigma} : L^1(G, M_n^*) \longrightarrow L^1(G, M_n^*)$, hence one can apply (i) and (ii) to $T_{\widetilde\sigma}$, noting that absolute continuity of σ is equivalent to that of $\widetilde\sigma$.

Finally, (viii) implies that each convolution operator $T_{\sigma_{ij}} : L^p(G) \longrightarrow L^p(G)$ is compact, by Lemma 2.1.4. Similar arguments as before show that $T_\sigma : L^p(G, M_n) \longrightarrow L^p(G, M_n)$ is compact for all $p \in [1, \infty]$, giving (vii). This concludes the proof.   □

3.3 Spectral Theory

In this section, we describe the spectrum of the convolution operator T_σ and study its eigenspaces.

Definition 3.3.1. Given a measure $\sigma \in M(G, M_n)$, we denote by Spec σ the spectrum of σ in the Banach algebra $M(G, M_n)$. By a previous remark, it is also the spectrum of σ regarded as an element in $M(G, M_n^*)$.

Given an operator $T : L^p(G, M_n) \longrightarrow L^p(G, M_n)$, we denote by $\mathrm{Spec}\,(T, L^p(G, M_n))$ and $\Lambda(T, L^p(G, M_n))$ its spectrum and the set of eigenvalues respectively. The continuous and residual spectra are denoted by $\mathrm{Spec}^c(T, L^p(G, M_n))$ and $\mathrm{Spec}^r(T, L^p(G, M_n))$ respectively. We note that, these spectra and eigenvalues remain unchanged if, in $L^p(G, M_n)$, we change the C*-norm on M_n to the trace norm or the Hilbert-Schmidt norm. As before, if no confusion is likely, we write $\mathrm{Spec}\,(T, L^p)$ for $\mathrm{Spec}\,(T, L^p(G, M_n))$ and $\Lambda(T, L^p)$ for $\Lambda(T, L^p(G, M_n))$.

Lemma 3.3.2. *Let $\sigma \in M(G, M_n)$. Then for $1 \leq p \leq \infty$ with conjugate exponent q, we have*

(i) $\Lambda(T_\sigma, L^1(G, M_n)) \subset \Lambda(T_\sigma, L^p(G, M_n)) \subset \Lambda(T_\sigma, L^\infty(G, M_n))$ *and they are equal*
   if G is compact,
(ii) $\mathrm{Spec}\,(T_\sigma, L^p(G, M_n))$
   $= \Lambda(T_\sigma, L^p(G, M_n)) \cup \Lambda(L_{\tilde{\sigma}}, L^q(G, M_n^*)) \cup \mathrm{Spec}^c(T_\sigma, L^p(G, M_n))$ for $p < \infty$,
(iii) $\mathrm{Spec}^c(T_\sigma, L^p(G, M_n)) = \mathrm{Spec}^c(L_{\tilde{\sigma}}, L^q(G, M_n^*))$ for $1 < p < \infty$,
(iv) $\mathrm{Spec}\,(T_\sigma, L^p(G, M_n)) \subset \mathrm{Spec}\,\sigma = \mathrm{Spec}(T_\sigma, L^1(G, M_n)) = \mathrm{Spec}(T_\sigma, L^\infty(G, M_n))$.

Proof. (i) The first inclusion is a simple consequence of $L^p * L^1 \subset L^1$. Let
$\alpha \in \Lambda(T_\sigma, L^p(G, M_n))$ and let $f \in L^p(G, M_n) \backslash \{0\}$ satisfy $f * \sigma = \alpha f$. Then by
Lemma 3.1.2, we can pick $h \in L^q(G, M_n)$ such that $\tilde{h} * f \in L^\infty(G, M_n) \backslash \{0\}$
and $(\tilde{h} * f) * \sigma = \alpha(\tilde{h} * f)$. So $\alpha \in \Lambda(T_\sigma, L^\infty)$. If G is compact, we have
$L^p(G, M_n) \subset L^1(G, M_n)$ for all p.
   (ii) This follows from the fact that

$$\mathrm{Spec}^r(T_\sigma, L^p) \subset \Lambda(T_\sigma^*, L^q) \subset \Lambda(T_\sigma, L^p) \cup \mathrm{Spec}^r(T_\sigma, L^p)$$

(cf. [24, p.581]).
   (iii) Let $1 < p < \infty$ and let $\alpha \in \mathrm{Spec}^c(T_\sigma, L^p)$. Then $\alpha \in \mathrm{Spec}(L_{\tilde{\sigma}}, L^q)$ and
$\overline{(T_\sigma - \alpha I)(L^p(G, M_n))} = L^p(G, M_n)$ implies that $L_{\tilde{\sigma}} - \alpha I : L^q(G, M^*) \longrightarrow L^q(G, M_n^*)$
is injective. Hence $\alpha \in \mathrm{Spec}^r(L_{\tilde{\sigma}}, L^q) \cup \mathrm{Spec}^c(L_{\tilde{\sigma}}, L^q)$. If $\alpha \in \mathrm{Spec}^r(L_{\tilde{\sigma}}, L^q)$, then
the proof of (ii) implies $\alpha \in \Lambda(T_\sigma, L^p)$ which is impossible. So $\alpha \in \mathrm{Spec}^c(L_{\tilde{\sigma}}, L^q)$.
Likewise $\mathrm{Spec}^c(L_{\tilde{\sigma}}, L^q) \subset \mathrm{Spec}^c(T_\sigma, L^p)$.
   (iv) If $\alpha \notin \mathrm{Spec}\,\sigma$, then $\sigma - \alpha\delta_e$ is invertible in $M(G, M_n)$ and hence $T_\sigma -$
$\alpha I : L^p(G, M_n) \longrightarrow L^p(G, M_n)$ is invertible with inverse $f \in L^p(G, M_n) \mapsto f *$
$(\sigma - \alpha\delta_e)^{-1} \in L^p(G, M_n)$. Therefore $\alpha \notin \mathrm{Spec}\,(T_\sigma, L^p(G, M_n))$.
   Next we show $\mathrm{Spec}\,\sigma \subset \mathrm{Spec}(T_\sigma, L^1)$. Let α be a complex number such that
$T_\sigma - \alpha I : L^1(G, M_n) \longrightarrow L^1(G, M_n)$ is invertible. We show that $\sigma - \alpha\delta_e$ is invertible
in $M(G, M_n)$. Let (ν_β) be an approximate identity in $L^1(G)$ weak* converging to the
unit mass $\delta_e \in M(G)$. Let

$$\bar{\nu}_\beta = \begin{pmatrix} \nu_\beta & & \\ & \ddots & \\ & & \nu_\beta \end{pmatrix} \in L^1(G, M_n).$$

Then $(\bar{\nu}_\beta)$ w*-converges to the identity $\delta_e \in M(G, M_n)$.
   Let $S : L^1(G, M_n) \longrightarrow L^1(G, M_n)$ be the inverse of $T_\sigma - \alpha I = T_{\sigma - \alpha\delta_e}$. Then
$S(\bar{\nu}_\beta) * (\sigma - \alpha\delta_e) = \bar{\nu}_\beta$. By choosing a subnet, we may assume that the net $(S(\bar{\nu}_\beta))$
weak* converges to some $\mu \in M(G, M_n)$. For each $h \in C_c(G, M_n)$, the net $(h * S(\bar{\nu}_\beta))$
in $L^\infty(G, M_n)$ weak* converges to $h * \mu$ since

$$\langle k, h * S(\bar{\nu}_\beta) \rangle = \langle \tilde{k} * h, \widetilde{S(\bar{\nu}_\beta)} \rangle \qquad (k \in C_c(G, M_n^*)).$$

Hence for every $f \in L^1(G, M_n^*)$, we have, by (3.4),

$$\langle f, h * S(\bar{v}_\beta) * (\sigma - \alpha\delta_e)\rangle = \langle (\tilde{\sigma} - \alpha\delta_e) *_\ell f, h * S(\bar{v}_\beta)\rangle$$
$$\longrightarrow \langle (\tilde{\sigma} - \alpha\delta_e) *_\ell f, h * \mu\rangle$$
$$= \langle f, h * \mu * (\sigma - \alpha\delta_e)\rangle$$

and also $\langle f, h * S(\bar{v}_\beta) * (\sigma - \alpha\delta_e)\rangle = \langle f, h * \bar{v}_\beta\rangle \longrightarrow \langle f, h * \delta_e\rangle$ which implies

$$\mu * (\sigma - \alpha\delta_e) = \delta_e$$

since $h \in C_c(G, M_n)$ was arbitrary. Thus $\sigma - \alpha\delta_e$ has a left inverse in $M(G, M_n)$.

It follows from $T_{\sigma - \alpha\delta_e} T_\mu = I$ that $S = T_\mu$ which gives $T_\mu T_{\sigma - \alpha\delta_e} = I$ and $(\sigma - \alpha\delta_e) * \mu = \delta_e$. Therefore $\sigma - \alpha\delta_e$ is invertible in $M(G, M_n)$.

Finally, we show $\mathrm{Spec}(L_{\tilde{\sigma}}, L^1(G, M_n)) = \mathrm{Spec}\,\sigma$ and note that

$$\mathrm{Spec}(T_{\tilde{\sigma}}, L^\infty(G, M_n)) = \mathrm{Spec}(L_{\tilde{\sigma}}, L^1(G, M_n)).$$

If $\alpha \notin \mathrm{Spec}\,\sigma$, then $L_{\tilde{\sigma}} - \alpha I = L_{\tilde{\sigma} - \alpha\delta_e}$ is invertible with inverse $L_{\tilde{\mu}}$ where μ is the inverse of $\sigma - \alpha\delta_e$ in $M(G, M_n)$ and we can use the formula $\widetilde{v * \mu} = \tilde{v} *_\ell \tilde{\mu}$ in (3.3).

Conversely, if $L_{\tilde{\sigma}} - \alpha I$ is invertible with inverse $S : L^1(G, M_n) \longrightarrow L^1(G, M_n)$, then $(\tilde{\sigma} - \alpha\delta_e) *_\ell S(\bar{v}_\beta) = \bar{v}_\beta$ where $(\bar{v}_\beta)$ is the above approximate identity and we may assume $S(\bar{v}_\beta))$ weak* converges to some $\mu \in M(G, M_n)$. For each $h \in C_c(G, M_n)$, the net $(S(\bar{v}_\beta) *_\ell h)$ weak* converges to $\mu *_\ell h$ since

$$\langle k, S(\bar{v}_\beta) *_\ell h\rangle = \langle \tilde{h} * k, S(\bar{v}_\beta)\rangle.$$

We have

$$\langle f, (\tilde{\sigma} - \alpha\delta_e) *_\ell \mu *_\ell h\rangle = \lim_\beta \langle f, (\tilde{\sigma} - \alpha\delta_e) *_\ell S(\bar{v}_\beta) *_\ell h\rangle = \langle f, \delta_e *_\ell h\rangle \quad (f \in L^1(G, M_n^*))$$

which implies $(\tilde{\sigma} - \alpha\delta_e) *_\ell \mu = \delta_e$ and $(\sigma - \alpha\delta_e) * \tilde{\mu} = \delta_e$. As before, $\sigma - \alpha\delta_e$ is then invertible in $M(G, M_n)$. $\square$

Example 3.3.3. The last inclusion in condition (i) in Lemma 3.3.2 is strict in general. Let σ be any *adapted* probability measure on a non-compact group G. By [10, Theorem 3.12], we have $1 \notin \Lambda(T_\sigma, L^p(G))$ for $1 \leq p < \infty$, but $1 \in \Lambda(T_\sigma, L^\infty(G))$.

Example 3.3.4. Condition (ii) in Lemma 3.3.2 does not hold for $p = \infty$ and condition (iii) does not hold for $p = 1, \infty$. Consider the Laplace operator $\Delta/2$ on the Euclidean space $\mathbb{R}^d$, which generates a convolution semigroup $\{\sigma_t\}_{t>0}$ of measures on $\mathbb{R}^d$:

$$d\sigma_t(x) = \frac{1}{(2\pi t)^{d/2}} \exp(-\|x\|^2/2t)dx.$$

We have $\tilde{\sigma}_t = \sigma_t$ and the convolution operator $T_{\sigma_t} : L^\infty(\mathbb{R}^d) \longrightarrow L^\infty(\mathbb{R}^d)$ is not weakly compact by Corollary 2.1.8. We have $\Lambda(T_{\sigma_t}, L^\infty) = \hat{\sigma}_t(\mathbb{R}^d) = (0, 1]$ (cf. Proposition 3.3.16), where

$$\hat{\sigma}_t(z) = \exp(-t\|z\|^2/2) \qquad (z \in \mathbb{R}^d)$$

is the Fourier transform of σ_t. Now $0 \in \mathrm{Spec}^r(T_{\sigma_t}, L^\infty(\mathbb{R}^d))$ since σ_t is absolutely continuous and we have $\overline{L^\infty(\mathbb{R}^d) * \sigma_t} \subset C(\mathbb{R}^d) \neq L^\infty(\mathbb{R}^d)$. Also $0 \in \mathrm{Spec}^c(T_{\widetilde{\sigma}_t}, L^1(\mathbb{R}^d))$ since $T_{\widetilde{\sigma}_t} : L^1(\mathbb{R}^d) \longrightarrow L^1(\mathbb{R}^d)$ has dense range by injectivity of T_{σ_t} on $L^\infty(\mathbb{R}^d)$.

A function $f \in L^\infty(\mathbb{R}^d)$ is an eigenfunction for $1 \in \Lambda(T_{\sigma_t}, L^\infty)$ for all $t > 0$ if, and only if, $\Delta f = 0$, in which case f is constant. By Example 3.3.3, $1 \notin \Lambda(T_{\sigma_t}, L^p(\mathbb{R}^d))$ for $1 \leq p < \infty$. In fact, the closure of Δ has no L^2 eigenfunction and $\Lambda(T_{\sigma_t}, L^2(\mathbb{R}^d)) = \emptyset$. However, we have $\mathrm{Spec}(T_{\sigma_t}, L^p(\mathbb{R}^d)) = [0,1]$ for $1 \leq p \leq \infty$ (cf. Theorem 3.3.23).

Example 3.3.5. The inclusion in (iv) in Lemma 3.3.2 is strict in general, for $p \in (1,\infty)$, even if G is abelian, by Remark 3.3.24. It has been shown by Sarnak [56] that for $\sigma \in M(\mathbb{T})$, where $\mathbb{T}$ is the circle group, one has $\mathrm{Spec}(T_\sigma, L^p(\mathbb{T})) = \mathrm{Spec}(T_\sigma, L^2(\mathbb{T}))$ for $1 < p < \infty$ if $\mathrm{Spec}(T_\sigma, L^2(\mathbb{T}))$ has zero capacity. We give a condition for $\mathrm{Spec}(T_\sigma, L^p(G)) = \mathrm{Spec}(T_\sigma, L^1(G))$ in Proposition 3.3.6.

Let $\alpha \notin \mathrm{Spec}(T_\sigma, L^p)$. Then the resolvent

$$R(\alpha, T_\sigma) = (\alpha I - T_\sigma)^{-1} : L^p(G, M_n) \longrightarrow L^p(G, M_n)$$

is M_n-linear and commutes with left translations. Indeed, for $A \in M_n$, $x \in G$ and $f \in L^p(G, M_n)$, we have

$$(\alpha - T_\sigma)(A R(\alpha, T_\sigma) f) = A(\alpha - T_\sigma) R(\alpha, T_\sigma) f = Af,$$
$$(\alpha - T_\sigma)\ell_x R(\alpha, T_\sigma) = \ell_x(\alpha - T_\sigma) R(\alpha, T_\sigma) = \ell_x.$$

Applying $R(\alpha, T_\sigma)$ to the left of both equations, we get $R(\alpha, T_\sigma)(Af) = A R(\alpha, T_\sigma) f$ and $\ell_x R(\alpha, T_\sigma) = R(\alpha, T_\sigma)\ell_x$. In particular, for $p = 1$, the resolvent $R(\alpha, T_\sigma)$ is a convolution operator by Corollary 3.1.11. This gives an alternative proof of $S = T_\mu$ in Lemma 3.3.2 (iv). For arbitrary p and $|\alpha| > \|\sigma\|$, we have

$$R(\alpha, T_\sigma) = \sum_{n=0}^{\infty} \frac{1}{\alpha^{n+1}} T_{\sigma^n}$$

and the series $\mu = \displaystyle\sum_{n=0}^{\infty} \frac{1}{\alpha^{n+1}} \sigma^n$ converges in $M(G, M_n)$ which gives $R(\alpha, T_\sigma) = T_\mu$.

For the L^p-spectrum to be identical with the L^1-spectrum, it is both necessary and sufficient that all resolvents be convolution operators.

Proposition 3.3.6. *Let $1 < p < \infty$ and $\sigma \in M(G, M_n)$. The following conditions are equivalent.*

(i) $\mathrm{Spec}(T_\sigma, L^p(G, M_n)) = \mathrm{Spec}(T_\sigma, L^1(G, M_n))$.
(ii) $R(\alpha, T_\sigma)$ *is a convolution operator for each* $\alpha \notin \mathrm{Spec}(T_\sigma, L^p)$.

Proof. (i) $\Longrightarrow$ (ii). Given $\alpha \notin \mathrm{Spec}(T_\sigma, L^p)$, we have $\alpha \notin \mathrm{Spec}\,\sigma$ and hence $(\sigma - \alpha \delta_e) * \mu = \mu * (\sigma - \alpha \delta_e) = \delta_e$ for some $\mu \in M(G, M_n)$. It follows that $(T_\sigma - \alpha I) T_\mu = T_\mu (T_\sigma - \alpha I) = I$ on $L^p(G, M_n)$ and hence $R(\alpha, T_\sigma) = T_\mu$.

(ii) $\implies$ (i). Let $\alpha \notin \mathrm{Spec}\,(T_\sigma, L^p(G, M_n))$. Then $R(\alpha, T_\sigma) = T_\mu$ for some $\mu \in M(G, M_n)$. We have $(T_\sigma - \alpha I)T_\mu = T_\mu(T_\sigma - \alpha I) = I$ on $L^1 \cap L^p$, and hence on L^1. Therefore $\alpha \notin \mathrm{Spec}\,(T_\sigma, L^1)$. $\square$

Let $\pi \in \widehat{G}$. Given $\sigma \in M(G, M_n)$ and $f \in L^1(G, M_n)$, we define their *Fourier transforms* by

$$\widehat{\sigma}(\pi) = \int_G (1_{M_n} \otimes \pi)(x^{-1})d(\sigma \otimes 1_{B(H_\pi)})(x) \in M_n \otimes \mathcal{B}(H_\pi)$$

and

$$\widehat{f}(\pi) = \int_G f(x) \otimes \pi(x^{-1})d\lambda(x)$$

$$= \begin{pmatrix} \int_G f_{11}(x)\pi(x^{-1})d\lambda(x) & \cdots & \int_G f_{1n}(x)\pi(x^{-1})d\lambda(x) \\ & \cdot & \\ & \cdot & \\ & \cdot & \\ \int_G f_{n1}(x)\pi(x^{-1})d\lambda(x) & \cdots & \int_G f_{nn}(x)\pi(x^{-1})d\lambda(x) \end{pmatrix} \in M_n \otimes \mathcal{B}(H_\pi)$$

where $M_n \otimes \mathcal{B}(H_\pi) = B(\mathbb{C}^n \otimes H_\pi)$ is a matrix algebra over $\mathcal{B}(H_\pi)$ and the Hilbert space tensor product $\mathbb{C}^n \otimes H_\pi$ identifies with the direct sum of n-copies of H_π. For $\iota \in \widehat{G}$, we have

$$\widehat{f}(\iota) = \int_G f d\lambda \in M_n.$$

In contrast to the scalar case, $\widehat{f * \sigma}(\pi)$ need not equal $\widehat{\sigma}(\pi)\widehat{f}(\pi)$. Instead, we have

$$\widehat{\sigma *_\ell f}(\pi) = \widehat{\sigma}(\pi)\widehat{f}(\pi)$$

although one still has $\widehat{f * \sigma} = \widehat{f}\widehat{\sigma}$ if G is abelian. However, if we define

$$\pi(\sigma) = \int_G (1_{M_n} \otimes \pi)(x)d(\sigma \otimes 1_{B(H_\pi)})(x) = \widehat{\widehat{\sigma}}(\pi) \tag{3.6}$$

and $\pi(f) = \int_G f(x) \otimes \pi(x)d\lambda(x)$, then, as in [9, p.36], we have

$$\pi(f * \sigma) = \pi(f)\pi(\sigma).$$

If $\widehat{f}(\pi) = 0$ for all $\pi \in \widehat{G}$, then $f = 0$. Indeed, we have

$$\int_G f_{ij}(x)\pi(x^{-1})d\lambda(x) = 0 \qquad (\pi \in \widehat{G})$$

which gives $f_{ij} = 0$ for all i, j. If G is abelian, then π is a character and we have

$$\widehat{\sigma}(\pi) = \int_G \pi(x^{-1})d\sigma(x) \in M_n$$

$$\widehat{f}(\pi) = \int_G f(x)\pi(x^{-1})d\lambda(x) \in M_n.$$

For abelian G, the inverse of $\pi \in \widehat{G}$ is given by

$$\widetilde{\pi}(x) = \pi(x^{-1}) \qquad (x \in G)$$

and we have $\widehat{\sigma}(\widetilde{\pi}) = \widehat{\widetilde{\sigma}}(\pi)$. Hence

$$\{\pi(\sigma) : \pi \in \widehat{G}\} = \{\widehat{\widetilde{\sigma}}(\pi) : \pi \in \widehat{G}\} = \{\widehat{\sigma}(\pi) : \pi \in \widehat{G}\}. \qquad (3.7)$$

Given $\mu \in M(G, M_n)$ and a function $f : G \longrightarrow M_n$, we define their *transposes* by pointwise transpose:

$$\mu^T(E) = \mu(E)^T, \quad f^T(x) = f(x)^T.$$

We note that

$$(\mu *_\ell f)^T = f^T * \mu^T.$$

For each $\pi \in \widehat{G}$, the Fourier transform

$$\widehat{\mu}(\pi) = \left(\int_G \pi(x^{-1})d\mu_{ij}(x) \right)$$

is a matrix of operators in $M_n \otimes B(H_\pi)$ and we have

$$\widehat{\mu^T}(\pi) = \left(\int_G \pi(x^{-1})d\mu_{ji}(x) \right) = \widehat{\mu}(\pi)^T.$$

It follows that, for each $\sigma \in M(G, M_n)$, we have

$$\operatorname{Spec} \widehat{\sigma}(\pi) = \operatorname{Spec} \widehat{\sigma^T}(\pi). \qquad (3.8)$$

Lemma 3.3.7. *Let $\sigma \in M(G, M_n)$ and $1 \leq p \leq \infty$. Then*

$$\operatorname{Spec}(T_\sigma, L^p(G, M_n)) = \operatorname{Spec}(L_{\sigma^T}, L^p(G, M_n)).$$

Proof. If $\alpha \in \mathbb{C}$ and $T_\sigma - \alpha I$ has an inverse $S \in B(L^p(G, M_n))$, we define $S^T \in B(L^p(G, M_n))$ by

$$S^T(f) = S(f)^T \qquad (f \in L^p(G, M_n)).$$

Then $L_{\sigma^T} - \alpha I$ has inverse S^T. Indeed, given $f \in L^p(G, M_n)$, we have

$$(L_{\sigma^T} - \alpha I)S^T(f) = \sigma^T *_\ell S(f)^T - \alpha S(f)^T = ((\sigma^T *_\ell S(f)^T)^T - \alpha S(f))^T$$
$$= S(f) * \sigma - \alpha S(f) = f = S^T(L_{\sigma^T} - \alpha I)(f).$$

The arguments can be reversed. $\qquad\qquad\qquad\qquad\qquad\qquad\qquad\qquad\qquad\qquad\qquad\square$

Proposition 3.3.8. *Let* $\sigma \in M(G,M_n)$. *Then*

$$\Lambda(T_\sigma,L^1) \subset \bigcup_{\pi\in\widehat{G}} \operatorname{Spec}\widehat{\sigma}(\pi) \subset \operatorname{Spec}\sigma.$$

Proof. Let $\alpha \in \Lambda(T_\sigma,L^1)$ with $f * \sigma = \alpha f$ for some nonzero $f \in L^1(G,M_n)$. Then there exists $\pi \in \widehat{G}$ such that $\widehat{f}(\pi) \neq 0$. We have $(\sigma^T *_\ell f^T)^T = f * \sigma = \alpha f$. Hence $(\sigma^T *_\ell f^T) = \alpha f^T$ and $\widehat{\sigma^T}(\pi)\widehat{f^T}(\pi) = \widehat{\sigma^T *_\ell f^T}(\pi) = \alpha\widehat{f^T}(\pi)$ which gives $(\widehat{\sigma^T}(\pi) - \alpha I)\widehat{f^T}(\pi) = 0$. Therefore $\widehat{\sigma^T}(\pi) - \alpha I$ is not invertible, that is, $\alpha \in \operatorname{Spec}\widehat{\sigma^T}(\pi)$.

If $\alpha \notin \operatorname{Spec}\sigma$, then $\sigma - \alpha\delta_e$ has an inverse $\mu \in M(G,M_n)$ and hence

$$(\sigma^T - \alpha\delta_e) *_\ell \mu^T = (\mu * (\sigma - \alpha\delta_e))^T = \delta_e = \mu^T *_\ell (\sigma^T - \alpha\delta_e).$$

It follows that, for all $\pi \in \widehat{G}$, we have $(\widehat{\sigma^T - \alpha\delta_e})(\pi)\widehat{\mu^T}(\pi) = I = \widehat{\mu^T}(\pi)$ $(\widehat{\sigma^T - \alpha\delta_e})(\pi)$, that is, $\widehat{\sigma^T}(\pi) - \alpha I$ is invertible in $M_n \otimes \mathcal{B}(H_\pi)$. This proves the last inclusion. $\qquad\square$

We first develop a matrix spectral theory for abelian groups, and consider non-abelian groups later. We need to extend the Plancherel theorem to the matrix setting. For this, denote by $M_{n,2}$ the vector space M_n equipped with the Hilbert-Schmidt norm and consider the Hilbert space $L^2(G,M_{n,2})$ with inner product $\langle\cdot,\cdot\rangle_2$, written as $\langle\cdot,\cdot\rangle$.

Lemma 3.3.9. (Plancherel Theorem) *Let* G *be abelian and let* $f \in L^1(G,M_{n,2}) \cap L^2(G,M_{n,2})$. *Then* $\|\widehat{f}\|_2 = \|f\|_2$ *and the mapping* $f \in L^1(G,M_{n,2}) \cap L^2(G,M_{n,2}) \mapsto \widehat{f} \in L^2(\widehat{G},M_{n,2})$ *extends to a unitary operator* $\mathcal{F} : L^2(G,M_{n,2}) \longrightarrow L^2(\widehat{G},M_{n,2})$.

Proof. The proof is similar to the scalar case, we outline the main steps which require matrix manipulation, for clarity. Let $f \in L^1(G,M_{n,2}) \cap L^2(G,M_{n,2})$ and let $f^* : G \longrightarrow M_n$ be the involution $f^*(x) = f(x^{-1})^*$. Then $f^* \in L^1(G,M_{n,2}) \cap L^2(G,M_{n,2})$ and $\widehat{f^*}(\gamma) = \widehat{f}(\gamma)^*$. The function $h = f * f^*$ belongs to $L^1(G,M_{n,2}) \cap L^2(G,M_{n,2})$ and has Fourier transform $\widehat{h} = \widehat{f}\widehat{f^*}$. Consider the scalar function $\operatorname{Tr}\circ h$ whose Fourier transform $\widehat{\operatorname{Tr}\circ h}$ equals $\operatorname{Tr}\circ\widehat{h} = \operatorname{Tr}\circ\widehat{f}\widehat{f^*}$ which is non-negative on $\widehat{G}$. By Lemma 3.1.1, $\operatorname{Tr}\circ h = \operatorname{Tr}\circ f * f^*$ is continuous. Hence, by the scalar result, $\widehat{\operatorname{Tr}\circ h} = \widehat{\operatorname{Tr}\circ h} \in L^1(\widehat{G})$ which gives $\widehat{h} \in L^1(G,M_{n,2})$ since $\|\widehat{h}(\gamma)\|_{hs} \leq \sqrt{n}\|\widehat{h}(\gamma)\|_{M_n} \leq \sqrt{n}\operatorname{Tr}(\widehat{h}(\gamma))$ by positivity of the matrx $\widehat{h}(\gamma)$. We also have

$$\operatorname{Tr}\circ h(e) = \int_{\widehat{G}} \widehat{\operatorname{Tr}\circ h}(\gamma)d\gamma$$

where $d\gamma$ is the Haar measure on $\widehat{G}$. It follows that

$$\|\widehat{f}\|_2^2 = \mathrm{Tr}\left(\int_{\widehat{G}} \widehat{f}(\gamma)\widehat{f}(\gamma)^* d\gamma\right) = \mathrm{Tr}(h(e))$$

$$= \mathrm{Tr}\left(\int_G f(y^{-1})f(y^{-1})^* d\lambda(y)\right) = \|f\|_2^2.$$

We can now extend the isometry $f \in L^1(G, M_{n,2}) \cap L^2(G, M_{n,2}) \mapsto \widehat{f} \in L^2(\widehat{G}, M_{n,2})$ to an isometry $\mathcal{F} : L^2(G, M_{n,2}) \longrightarrow L^2(\widehat{G}, M_{n,2})$ and it remains to show that $\mathcal{F}$ is surjective. Indeed, if $g \in L^2(\widehat{G}, M_{n,2})$ satisfies $\langle \mathrm{Im}\mathcal{F}, g\rangle = 0$, then we have, for each $\varphi \in L^2(G, M_{n,2})$,

$$\langle \mathcal{F}'(g), \varphi\rangle = \mathrm{Tr}\left(\int_G \varphi^* \mathcal{F}'(g)\right) = \mathrm{Tr}\left(\int_{\widehat{G}} \mathcal{F}(\varphi^*)g\right) = 0$$

where the map $\mathcal{F}' : L^2(\widehat{G}, M_{n,2}) \longrightarrow L^2(G, M_{n,2})$ is constructed similarly. Hence $\mathcal{F}'(g) = 0$ in $L^2(G, M_{n,2})$ and $g = 0$ in $L^2(\widehat{G}, M_{n,2})$.   □

Given a subset $\mathcal{E} \subset M_n$, we denote by

$$\Lambda\mathcal{E} = \{\alpha \in \mathbb{C} : \det(A - \alpha I) = 0 \text{ for some } A \in \mathcal{E}\}$$

the set of all eigenvalues of the matrices in $\mathcal{E}$. Using Lemma 3.3.9, the L^2-spectrum of T_σ on an abelian group can be determined without difficulty. We prove a lemma first.

Lemma 3.3.10. *Let Ω be a locally compact Hausdorff space and let f be an element in the algebra $C_b(\Omega, M_n)$ of bounded continuous M_n-valued functions on Ω. Then the spectrum of f in this algebra is given by*

$$\mathrm{Spec} f = \overline{\Lambda\{f(\omega) : \omega \in \Omega\}}.$$

Proof. Since $C_b(\Omega, M_n) = C(\overline{\Omega}, M_n)$, where $\overline{\Omega}$ is the Stone-Čech compactification, it is straightforward to show, using determinant, that

$$\mathrm{Spec} f = \Lambda\{f(\omega) : \omega \in \overline{\Omega}\} \supset \overline{\Lambda\{f(\omega) : \omega \in \Omega\}}.$$

To see the reverse inclusion, let α be an eigenvalue of $f(\omega)$ for some $\omega \in \overline{\Omega}$ with $\omega = \lim_\beta \omega_\beta$ and $\omega_\beta \in \Omega$. Then we have $\lim_\beta \det(f(\omega_\beta) - \alpha) = \det(f(\omega) - \alpha) = 0$. If $\det(f(\omega_\beta) - \alpha) = 0$ for some β, there is nothing to prove. Otherwise, let $\varepsilon > 0$ and choose $|\det(f(\omega_\beta) - \alpha)| < \varepsilon^n$. Let ξ be the eigenvalue of $f(\omega_\beta) - \alpha I$ with the least modulus. Since determinant is the product of eigenvalues, we have $|\xi| < \varepsilon$. Now $\alpha + \xi$ is an eigenvalue of $f(\omega_\beta)$ and $|(\alpha + \xi) - \alpha| = |\xi| < \varepsilon$. This proves that α is in the closure of $\Lambda\{f(\omega) : \omega \in \Omega\}$.   □

Remark 3.3.11. The above arguments also show that, if f vanishes at infinity, then we have $0 \in \overline{\Lambda\{f(\omega) : \omega \in \Omega\}}$.

Proposition 3.3.12. *Let G be abelian and $\sigma \in M(G, M_n)$. Then the convolution operator $L_\sigma : L^2(G, M_{n,2}) \longrightarrow L^2(G, M_{n,2})$ is unitarily equivalent to the multiplication operator $M_{\widehat{\sigma}} : h \in L^2(\widehat{G}, M_{n,2}) \mapsto \widehat{\sigma} h \in L^2(\widehat{G}, M_{n,2})$ via the Fourier transform $\mathcal{F} : L^2(G, M_{n,2}) \longrightarrow L^2(\widehat{G}, M_{n,2})$, that is,*

$$\mathcal{F} L_\sigma = M_{\widehat{\sigma}} \mathcal{F}$$

and we have $\mathrm{Spec}(T_\sigma, L^2(G, M_n)) = \overline{\Lambda\{\widehat{\sigma}(\gamma) : \gamma \in \widehat{G}\}}$.

Proof. Define a map $\psi : L^\infty(\widehat{G}, M_n) \longrightarrow B(L^2(G, M_{n,2}))$ by

$$\psi(f)(h) = \mathcal{F}^{-1}(f\mathcal{F}(h)) \qquad (f \in L^\infty(\widehat{G}, M_{n,2}), h \in L^2(G, M_{n,2})).$$

One can verify readily that ψ is injective and also an algebra homomorphism. Moreover, ψ is a *-homomorphism. Indeed, given $f \in L^\infty(\widehat{G}, M_n)$ and $h, k \in L^2(G, M_{n,2})$, we have

$$
\begin{aligned}
\langle k, \psi(f^*)h \rangle &= \langle k, \mathcal{F}^{-1}(f^*\mathcal{F}(h)) \rangle = \langle \mathcal{F}(k), f^*\mathcal{F}(h) \rangle \\
&= Tr\left(\int_G \mathcal{F}(k)(f^*\mathcal{F}(h))^* d\lambda \right) = Tr\left(\int_G \mathcal{F}(k)\mathcal{F}(h)^* f d\lambda \right) \\
&= Tr\left(\int_G f\mathcal{F}(k)\mathcal{F}(h)^* d\lambda \right) = \langle f\mathcal{F}(k), \mathcal{F}(h) \rangle \\
&= \langle \mathcal{F}^{-1}(f\mathcal{F}(k)), h \rangle.
\end{aligned}
$$

Hence ψ is an isometry (cf. [62, Corollary I.5.4]) and $L^\infty(\widehat{G}, M_n)$ identifies as a unital C*-subalgebra of $B(L^2(G, M_{n,2}))$. We have $\widehat{\sigma} \in L^\infty(\widehat{G}, M_n)$ and

$$\psi(\widehat{\sigma})(h) = \mathcal{F}^{-1}(\widehat{\sigma}\mathcal{F}(h)) = \mathcal{F}^{-1}(\mathcal{F}(\sigma *_\ell h)) = \sigma *_\ell h = L_\sigma(h) \qquad (h \in L^2(G, M_{n,2}))$$

that is, $\psi(\widehat{\sigma}) = L_\sigma$ and $\mathcal{F} L_\sigma = M_{\widehat{\sigma}} \mathcal{F}$. Also we have

$$\mathrm{Spec}(L_\sigma, L^2) = \mathrm{Spec}_{B(L^2(G, M_{n,2}))} \psi(\widehat{\sigma}) = \mathrm{Spec}_{L^\infty(\widehat{G}, M_n)} \widehat{\sigma}$$

where the second equality follows from that $L^\infty(\widehat{G}, M_n)$ is a unital C*-subalgebra of $B(L^2(G, M_{n,2}))$ (cf. [62, Proposition I.4.8]). Since $T_{\widehat{\sigma}}^* = L_\sigma$, we obtain

$$\mathrm{Spec}(T_{\widetilde{\sigma}}, L^2) = \mathrm{Spec}(T_{\widehat{\sigma}}^*, L^2) = \mathrm{Spec}(L_\sigma, L^2) = \mathrm{Spec}_{L^\infty(\widehat{G}, M_n)} \widehat{\sigma}$$

and hence, noting that $C_b(\widehat{G}, M_n)$ is a unital C*-subalgebra of $L^\infty(\widehat{G}, M_n)$,

$$\mathrm{Spec}(T_\sigma, L^2) = \mathrm{Spec}_{L^\infty(\widehat{G}, M_n)} \widetilde{\widehat{\sigma}} = \mathrm{Spec}_{C_b(\widehat{G}, M_n)} \widetilde{\widehat{\sigma}}$$

which is the closure of $\Lambda\{\widehat{\sigma}(\gamma) : \gamma \in \widehat{G}\}$ by Lemma 3.3.10 and (3.7). $\qquad\square$

To define Fourier transform for $1 < p < 2$, we use the Hausdorff-Young inequality as in the scalar case.

Lemma 3.3.13. (Hausdorff-Young) *Let G be an abelian group and let $1 < p < 2$. Given $f \in L^p(G,M_n) \cap L^1(G,M_n)$, we have $\widehat{f} \in L^q(\widehat{G},M_n)$ and $\|\widehat{f}\|_q \leq n^2 \|f\|_p$.*

Proof. Let μ be the Haar measure on $\widehat{G}$ such that we have the Hausdorff-Young inequality

$$\|\widehat{h}\|_q \leq \|h\|_p$$

for $h \in L^p(G) \cap L^1(G)$ (cf. [27, 4.27]). Let $f = (f_{ij}) \in L^p(G,M_n) \cap L^1(G,M_n)$. For $\pi \in \widehat{G}$, the Fourier transform $\widehat{f}(\pi)$ is the matrix $(\widehat{f_{ij}}(\pi)) \in M_n$ and we have

$$\left(\int_{\widehat{G}} \|\widehat{f}(\pi)\|_{M_n}^q d\mu(\pi) \right)^{1/q} \leq \left(\int_{\widehat{G}} \left(\sum_{ij} |\widehat{f_{ij}}(\pi)|^2 \right)^{q/2} d\mu(\pi) \right)^{1/q}$$

$$\leq \left(\int_{\widehat{G}} \left(\sum_{ij} |\widehat{f_{ij}}(\pi)| \right)^q d\mu(\pi) \right)^{1/q}$$

$$\leq \sum_{ij} \|\widehat{f_{ij}}\|_q \leq n^2 \|f\|_p.$$

$\square$

Since $L^p(G,M_n) \cap L^1(G,M_n)$ is $\|\cdot\|_p$-dense in $L^p(G,M_n)$, the above lemma implies that, for $1 < p < 2$, the Fourier transform $f \in L^p(G,M_n) \cap L^1(G,M_n) \mapsto \widehat{f} \in L^q(G,M_n)$ has a unique extension to a bounded linear operator on $L^p(G,M_n)$, which will still be denoted by $f \mapsto \widehat{f}$ and called the *Fourier transform* of $L^p(G,M_n)$.

To obtain further spectral results for abelian groups G, we introduce a useful device, namely, the determinant of a matrix-valued measure. Let $\sigma = (\sigma_{ij}) \in M(G,M_n)$ where G is abelian. We define its *determinant*, $\det \sigma$, which is a complex-valued measure, by convolution:

$$\det \sigma = \sum_{\tau} \operatorname{sgn}(\tau) \sigma_{1\tau(1)} * \cdots * \sigma_{n\tau(n)}$$

where τ is a permutation of $\{1, \ldots, n\}$. This is well-defined since G is abelian. We can now define the *adjugate matrix* of σ, $Adj\,\sigma \in M(G,M_n)$, by convolution such that

$$(Adj\,\sigma) * \sigma = \sigma * (Adj\,\sigma) = \begin{pmatrix} \det \sigma & & \\ & \ddots & \\ & & \det \sigma \end{pmatrix}.$$

Given that G is abelian, $\widehat{G}$ is the group of characters and we have

$$\widehat{\det \sigma}(\pi) = \det \widehat{\sigma}(\pi) \qquad (\pi \in \widehat{G}).$$

We have the following matrix version of a Tauberian theorem.

Lemma 3.3.14. *Let G be abelian and let* $\sigma \in M(G, M_n)$. *Then the following conditions are equivalent.*

(i) *For each* $f \in L^\infty(G, M_n)$, $f * \sigma = 0$ *implies* f *is constant.*
(ii) *For each* $f \in L^\infty(G, M_n^*)$, $\sigma *_\ell f = 0$ *implies* f *is constant.*
(iii) $\det \widehat{\sigma}(\pi) \neq 0$ *for every* $\pi \in \widehat{G}\backslash\{\iota\}$.

Proof. The equivalence of (i) and (iii) has been proved in [12].

(ii) $\Longrightarrow$ (iii). Suppose $\det \widehat{\sigma}(\pi) = 0$ for some $\pi \neq \iota$. We can find a non-zero vector $\xi = (\xi_1, \ldots, \xi_n) \in \mathbb{C}^n$ such that $\widehat{\sigma}(\pi)\xi = 0$. Define a function $f \in L^\infty(G, M_n^*)$ by

$$f(x) = \pi(x) \begin{pmatrix} \xi_1 & \cdots & \xi_1 \\ \vdots & & \vdots \\ \xi_n & \cdots & \xi_n \end{pmatrix}.$$

We have

$$\sigma *_\ell f(x) = \int_G d\sigma(y) f(xy^{-1})$$

$$= \pi(x) \int_G \pi(y^{-1}) d\sigma(y) \begin{pmatrix} \xi_1 & \cdots & \xi_1 \\ \vdots & & \vdots \\ \xi_n & \cdots & \xi_n \end{pmatrix} = 0$$

but f is non-constant.

(iii) $\Longrightarrow$ (ii). Consider the complex measure $\det \sigma$ whose Fourier transform satisfies $\widehat{\det \sigma}(\pi) \neq 0$ for each $\pi \in \widehat{G}\backslash\{\iota\}$. Let $f = (f_{ij}) \in L^\infty(G, M_n^*)$ be such that $\sigma *_\ell f = 0$. Let f^T and σ^T be the transposes defined pointwise. Then

$$f^T * \sigma^T(x) = \int_G f^T(xy^{-1}) d\sigma^T(y) = (\sigma *_\ell f)(x)^T = 0$$

for λ-a.e. $x \in G$. Therefore

$$f^T * \begin{pmatrix} \det \sigma^T & & \\ & \ddots & \\ & & \det \sigma^T \end{pmatrix} = f^T * \sigma^T * Adj\,\sigma^T = 0$$

which gives $f_{ij}^T * \det \sigma = 0$ where $f_{ij}^T = f_{ji} \in L^\infty(G)$. We can apply the equivalence (i) $\Leftrightarrow$ (iii) to $\det \sigma$ and $L^\infty(G)$. This implies that f_{ji} is constant for all i, j. Hence f is constant. $\qquad \square$

Remark 3.3.15. In the above lemma, (i) $\Longrightarrow$ (iii) can be extended to non-abelian groups in which case, (i) or (ii) implies that 0 is not an eigenvalue of $\widehat{\sigma}(\pi)$ for each $\pi \in \widehat{G}\backslash\{\iota\}$. However, (iii) $\Longrightarrow$ (i) fails for non-abelian groups. Let σ be an adapted probability measure on a non-amenable group G. Then there exists a *non-constant* function $f \in L^\infty(G)$ such that $f * \sigma = f$ (cf. [13, Proposition 2.1.3]). By [10,

Lemma 3.11], $\widehat{\sigma - \delta_e}(\pi)$ is invertible in $\mathcal{B}(H_\pi)$ for every $\pi \in \widehat{G} \setminus \{\iota\}$ while f satisfies $f * (\sigma - \delta_e) = 0$. This reveals a different spectral phenomenon between abelian and non-abelian groups.

Proposition 3.3.16. *Let G be abelian and let $\sigma \in M(G, M_n)$. Then*

$$\Lambda(T_\sigma, L^\infty(G, M_n)) = \Lambda\{\widehat{\sigma}(\pi) : \pi \in \widehat{G}\}$$

and $\Lambda(T_\sigma, L^p) \subset \Lambda \widehat{\sigma}(\widehat{G})$ for all $p < \infty$.

Proof. Let α be an eigenvalue of T_σ so that $f * \sigma = \alpha f$ for some non-zero $f \in L^\infty(G, M_n)$. If f is constant and equal $A \in M_n$ say, then $\alpha A = A\sigma(G) = A\widehat{\sigma}(\iota)$ gives $A(\widehat{\sigma}(\iota) - \alpha I) = 0$. Hence $\widehat{\sigma}(\iota) - \alpha I$ is not invertible since $A \neq 0$, that is, α is an eigenvalue of $\widehat{\sigma}(\iota)$.

If f is non-constant, then $f * (\sigma - \alpha \delta_e) = 0$ and Lemma 3.3.14 imply that there is some $\pi \in \widehat{G}$ such that $\det \widehat{\sigma - \alpha \delta_e}(\pi) = 0$. Hence

$$\det(\widehat{\sigma}(\pi) - \alpha I) = \det \widehat{\sigma - \alpha \delta_e}(\pi) = 0$$

that is, α is an eigenvalue of $\widehat{\sigma}(\pi)$.

Let $\alpha \in \Lambda\{\widehat{\sigma}(\pi) : \pi \in \widehat{G}\}$ and say, α is an eigenvalue of $\widehat{\sigma}(\pi)$. Then there exists a non-zero vector $\xi = (\xi_1, \ldots, \xi_n) \in \mathbb{C}^n$ such that $\widehat{\sigma}(\pi)^* \xi = \bar{\alpha} \xi$. Define a non-zero function $f \in L^\infty(G, M_n)$ by

$$f(x) = \pi(x) \begin{pmatrix} \bar{\xi}_1 & \cdots & \bar{\xi}_n \\ 0 & \cdots & 0 \\ \vdots & & \vdots \\ 0 & \cdots & 0 \end{pmatrix} \qquad (x \in G).$$

Then we have

$$f * \sigma(x) = \pi(x) \begin{pmatrix} \bar{\xi}_1 & \cdots & \bar{\xi}_n \\ 0 & \cdots & 0 \\ \vdots & & \vdots \\ 0 & \cdots & 0 \end{pmatrix} \widehat{\sigma}(\pi) = \alpha \pi(x) \begin{pmatrix} \bar{\xi}_1 & \cdots & \bar{\xi}_n \\ 0 & \cdots & 0 \\ \vdots & & \vdots \\ 0 & \cdots & 0 \end{pmatrix}$$

since

$$\widehat{\sigma}(\pi)^* \begin{pmatrix} \xi_1 & 0 & \cdots & 0 \\ \vdots & \vdots & & \vdots \\ \xi_n & 0 & \cdots & 0 \end{pmatrix} = \bar{\alpha} \begin{pmatrix} \xi_1 & 0 & \cdots & 0 \\ \vdots & \vdots & & \vdots \\ \xi_n & 0 & \cdots & 0 \end{pmatrix}.$$

Hence $f * \sigma = \alpha f$ and $\alpha \in \Lambda(T_\sigma, L^\infty)$.

The last assertion follows from Lemma 3.3.2 (i). $\square$

Proposition 3.3.17. *Let G be abelian and let $\sigma \in M(G, M_n)$. Then we have*

$$\Lambda\{\widehat{\sigma}(\pi) : \pi \in \widehat{G}\} \subset \mathrm{Spec}(T_\sigma, L^p)$$

for all $p \in [1, \infty]$.

Proof. First, consider the case $1 < p < 2$. Let α be an eigenvalue of $\widehat{\sigma}(\pi)$ for some $\pi \in \widehat{G}$. We show that

$$T_\sigma - \alpha I : L^p(G, M_n) \longrightarrow L^p(G, M_n)$$

is not invertible. Suppose otherwise, it has a bounded inverse $S : L^p(G, M_n) \longrightarrow L^p(G, M_n)$ which commutes with left translations. Let K be a compact subset of G with positive Haar measure and define the function $k \in L^p(G, M_n) \cap L^2(G, M_n)$ by

$$k = \begin{pmatrix} \pi\chi_K & & \\ & \ddots & \\ & & \pi\chi_K \end{pmatrix}$$

where χ_K is the characteristic function of K. By surjectivity, there exits $h \in L^p(G, M_n)$ such that

$$h * \sigma - \alpha h = k$$

which, as G is abelian, gives $\widehat{h}\widehat{\sigma} - \alpha\widehat{h} = \widehat{k}$ almost everywhere on $\widehat{G}$. Let $D : L^p(G) \longrightarrow L^p(G, M_n)$ be the natural embedding

$$D(f) = \begin{pmatrix} f & & \\ & \ddots & \\ & & f \end{pmatrix}.$$

Given $1 \leq i, j \leq n$, let $\psi : L^p(G, M_n) \longrightarrow L^p(G)$ be the projection onto the ij-component: $\psi(g) = g_{ij}$ for $g = (g_{ij})$. Then the mapping $\psi \circ S \circ D : L^p(G) \longrightarrow L^p(G)$ commutes with left translations and by [44, Corollary 4.1.2], there exists a function $\varphi \in L^\infty(\widehat{G})$ such that

$$\widehat{h}_{ij} = \widehat{\psi(h)} = \widehat{\psi SD(\pi\chi_K)} = \varphi\widehat{\pi\chi_K} \in L^\infty(\widehat{G}).$$

It follows that $\widehat{h} \in L^\infty(\widehat{G}, M_n)$.

Let $\{V_\beta\}$ be a net of neighbourhoods of π decreasing to $\{\pi\}$. Since V_β has positive Haar measure, we can find $\gamma_\beta \in V_\beta$ such that $\|\widehat{h}(\gamma_\beta)\|_{M_n} \leq \|\widehat{h}\|_\infty$ and

$$\widehat{h}(\gamma_\beta)\widehat{\sigma}(\gamma_\beta) - \alpha\widehat{h}(\gamma_\beta) = \widehat{k}(\gamma_\beta).$$

Therefore

$$|\det\widehat{k}(\gamma_\beta)| \leq \|\widehat{h}(\gamma_\beta)\|_{M_n}^n |\det(\widehat{\sigma}(\gamma_\beta) - \alpha)| \leq \|h\|_\infty^n |\det(\widehat{\sigma}(\gamma_\beta) - \alpha)|$$

where $\widehat{k}$ is continuous, giving

$$0 < \lambda(K)^n = |\det \widehat{k}(\pi)| \leq \|\widehat{h}\|_\infty^n |\det(\widehat{\sigma}(\pi) - \alpha)| = 0$$

which is a contradiction. Therefore $T_\sigma - \alpha I$ is not invertible and $\alpha \in \mathrm{Spec}(T_\sigma, L^p)$.

For $2 < p < \infty$, the conjugate exponent q satisfies $1 < q < 2$ and the above arguments are applicable to $L_{\widetilde{\sigma}}$ which commutes with left translations. Hence we have

$$\mathrm{Spec}(T_\sigma, L^p) = \mathrm{Spec}(L_{\widetilde{\sigma}}, L^q) \supset \Lambda \widehat{\widetilde{\sigma}}(\widehat{G}) = \Lambda \widehat{\sigma}(\widehat{G}).$$

$$\square$$

The reverse inclusion for the above result requires absolute continuity of σ. To prove it, we use a Wiener-Levy type theorem and we first develop some technical tools by adapting the ideas for the scalar case in [55] to the matrix setting.

Let G be an abelian group. We begin by noting that, given a compact set K contained in an open set W in the dual group $\widehat{G}$, one can find a function $f \in L^1(G, M_n)$ such that

$$\widehat{f} = \begin{cases} I_{M_n} & \text{on } K \\ 0 & \text{on } \widehat{G} \backslash W. \end{cases}$$

Indeed, one can find $h \in L^1(G)$ whose Fourier transform $\widehat{h}$ equals 1 on K, and vanishes outside W. Then the diagonal

$$f = D(h) = \begin{pmatrix} h & & \\ & \ddots & \\ & & h \end{pmatrix}$$

satisfies the requirements. Let

$$A(\widehat{G}, M_n) = \{\widehat{f} : f \in L^1(G, M_n)\}.$$

Then, as in the scalar case, $A(\widehat{G}, M_n)$ is a Banach algebra under the pointwise product and the norm

$$\||\widehat{f}\|| := \|f\|_1.$$

Naturally, we call $A(\widehat{G}, M_n)$ the *matrix Fourier algebra* of $\widehat{G}$.

Lemma 3.3.18. *Let G be abelian and let $f \in L^1(G, M_n)$. Given $\zeta \in \widehat{G}$ with a neighbourhood $W \subset \widehat{G}$ of ζ, and given $\varepsilon > 0$, there exists a function $h \in L^1(G, M_n)$ with $\|h\|_1 < \varepsilon$ such that $\widehat{h} = 0$ on $\widehat{G} \backslash W$ and*

$$\widehat{f}(\gamma) - \widehat{h}(\gamma) = \widehat{f}(\zeta)$$

in some neighbourhood of ζ.

Proof. This follows easily from the scalar result. Let $f = (f_{ij})$. By [55, Theorem 2.6.5], we can find $h_{ij} \in L^1(G)$ with $\|h_{ij}\|_1 < \varepsilon/n^2$ such that $\widehat{h_{ij}} = 0$ outside W and $\widehat{f_{ij}}(\gamma) - \widehat{h_{ij}}(\gamma) = \widehat{f_{ij}}(\zeta)$ in some neighbourhood V_{ij} of ζ. Let $h = (h_{ij}) \in L^1(G, M_n)$. Then

$$\|h\|_1 = \int_G \|h(x)\|_{M_n} d\lambda(x) \leq \int_G \left(\sum_{i,j} |h_{ij}(x)|^2 \right)^{1/2} d\lambda(x)$$

$$\leq \sum_{i,j} \int_G |h_{ij}(x)| d\lambda(x) < n^2 \left(\frac{\varepsilon}{n^2} \right) = \varepsilon$$

and we have $\widehat{h} = (\widehat{h_{ij}}) = 0$ outside W with $\widehat{f}(\gamma) - \widehat{h}(\gamma) = \widehat{f}(\zeta)$ in the neighbourhood $\bigcap_{i,j} V_{ij}$ of ζ.   □

Lemma 3.3.19. *Let G be abelian and let $f \in L^1(G, M_n)$. Given $\varepsilon > 0$, there exists a function $v \in L^1(G, M_n)$ such that its Fourier transform $\widehat{v}$ has compact support and $\|f - f * v\|_1 < \varepsilon$.*

Proof. We assume $f \neq 0$. We first note that the set

$$\{v \in L^1(G, M_n) : \widehat{v} \text{ has compact support}\}$$

is dense in $L^1(G, M_n)$. Indeed, by the scalar result, given $f = (f_{ij}) \in L^1(G, M_n)$ and for $\delta > 0$, one can find $v = (v_{ij}) \in L^1(G, M_n)$ satisfying

$$\|v_{ij} - f_{ij}\|_1 < \frac{\delta}{n^2}$$

and $\widehat{v_{ij}}$ has compact support. Since the support of $\widehat{v}$ is contained in $\bigcup_{i,j} \operatorname{supp} \widehat{v_{ij}}$, it is also compact. As before, we have

$$\|v - f\|_1 \leq \sum_{i,j} \|v_{ij} - f_{ij}\|_1 < n^2 \left(\frac{\delta}{n^2} \right) = \delta.$$

Now, since $L^1(G, M_n)$ has an approximate identity, we can find $u \in L^1(G, M_n)$ such that $\|f - f * u\|_1 < \varepsilon/2$. Choose $v \in L^1(G, M_n)$ such that $\widehat{v}$ has compact support and

$$\|u - v\| < \frac{\varepsilon}{2\|f\|_1}.$$

Then we have

$$\|f - f * v\|_1 \leq \|f - f * u\|_1 + \|f * (u - v)\|_1 < \varepsilon.$$

□

Definition 3.3.20. Let G be an abelian group. A function $\psi : \widehat{G} \longrightarrow M_n$ is said to *belong to* $A(\widehat{G}, M_n)$ *locally at* $\zeta \in \widehat{G}$ if there are a neighbourhood V of ζ and a function $\widehat{f} \in A(\widehat{G}, M_n)$ such that $\psi = \widehat{f}$ on V. If $\widehat{G}$ is not compact, we say that ψ *belongs to* $A(\widehat{G}, M_n)$ *at* ∞ if there are a compact set $K \subset \widehat{G}$ and a function $\widehat{f} \in A(\widehat{G}, M_n)$ such that $\psi = \widehat{f}$ on $\widehat{G} \backslash K$.

Lemma 3.3.21. *If a function* $\psi : \widehat{G} \longrightarrow M_n$ *belongs to* $A(\widehat{G}, M_n)$ *locally at every point of* $\widehat{G}$, *including* ∞ *if* $\widehat{G}$ *is not compact, then we have* $\psi \in A(\widehat{G}, M_n)$.

Proof. First suppose ψ has compact support $K \subset \widehat{G}$. Then there are open sets $V_1, \ldots, V_k$ and functions $\widehat{f}_1, \ldots \widehat{f}_k \in A(\widehat{G}, M_n)$ such that $K \subset V_1 \cup \cdots \cup V_k$ and $\psi = \widehat{f}_i$ on V_i. Choose open sets $W_i \subset V_i$ with compact closure $\overline{W}_i \subset V_i$ and $K \subset W_i \cup \cdots \cup W_k$. As noted earlier, we can find $\widehat{h}_i \in A(\widehat{G}, M_n)$ satisfying

$$\widehat{h}_i = \begin{cases} I \text{ on } \overline{W}_i \\ 0 \text{ on } \widehat{G} \backslash V_i. \end{cases}$$

We have $\psi \widehat{h}_i = \widehat{f}_i \widehat{h}_i \in A(\widehat{G}, M_n)$ for each i which implies

$$\psi = \psi(1 - (1 - \widehat{h}_1)(1 - \widehat{h}_2) \cdots (1 - \widehat{h}_k)) \in A(\widehat{G}, M_n).$$

Now, without any assumption on ψ, since ψ belongs to $A(\widehat{G}, M_n)$ at ∞, there are a compact set $K \subset \widehat{G}$ and a function $\widehat{g} \in A(\widehat{G}, M_n)$ such that $\psi = \widehat{g}$ outside K. Therefore $\psi - \widehat{g}$ has compact support and by the above arguments, we have $\psi - \widehat{g} \in A(\widehat{G}, M_n)$ and hence $\psi \in A(\widehat{G}, M_n)$. $\square$

A holomorphic map $F : E \longrightarrow F$ between complex Banach spaces has a series expansion around a point $z_0 \in E$ of the form

$$F(z) = \sum_{n=0}^{\infty} \frac{1}{n!} D^n F(z_0)(z - z_0, \ldots, z - z_0)$$

where $D^n F(z_0) : E^n \longrightarrow F$ is the n-th derivative of F at z_0. We will use another form of the Taylor series for a holomorphic map $F : M_n \longrightarrow M_n$ which is more suitable to our purpose. We note that, when equipped with the Hilbert-Schmidt norm $\| \cdot \|_{hs}$, the Hilbert space M_n identifies with the complex Euclidean space $\mathbb{C}^{n^2}$ via

$$(z_{ij}) \in M_n \mapsto (z_{11}, \ldots, z_{n1}, z_{12}, \ldots, z_{n2}, \ldots, z_{nn}) \in \mathbb{C}^{n^2}.$$

Therefore, by considering each ij-th entry of

$$F = (F_{ij}) : \mathbb{C}^{n^2} \longrightarrow M_n,$$

its Taylor series near $w = (w_{ij})$ can be written in the form

$$F(z) = \sum_{\kappa} A_{\kappa}(z - w)^{\kappa}$$

where $A_\kappa \in M_n$ and we adopt the usual convention of multi-indices: $\kappa = (\kappa_{11}, \ldots, \kappa_{nn})$ and $w^\kappa = w_{11}^{\kappa_{11}} \cdots w_{nn}^{\kappa_{nn}}$. We note that the neighbourhoods of w can be described by any norm on M_n since all norms are equivalent on M_n.

We are now ready to derive a matrix version of a Wiener-Levy type theorem. We note that each $\widehat{f} \in A(\widehat{G}, M_n)$ vanishes at infinity and therefore the closure of its range in M_n contains 0 if $\widehat{G}$ is not compact.

Theorem 3.3.22. *Let* $\widehat{f} \in A(\widehat{G}, M_n)$ *and let* $\mathcal{U}$ *be an open subset of* M_n *containing the closure* $\overline{\widehat{f}(\widehat{G})}$. *Then for any holomorphic map* $F : \mathcal{U} \longrightarrow M_n$ *satisfying* $F(0) = 0$ *if* $\widehat{G}$ *is non-compact, there exists a function* $\widehat{\varphi} \in A(\widehat{G}, M_n)$ *such that*

$$\widehat{\varphi}(\gamma) = F(\widehat{f}(\gamma)) \qquad (\gamma \in \widehat{G}).$$

We will denote $\widehat{\varphi}$ *by* $F(\widehat{f})$.

Proof. We need to show that the function $F \circ \widehat{f} : \widehat{G} \longrightarrow M_n$ belongs to $A(\widehat{G}, M_n)$. By Lemma 3.3.21, it suffices to show that $F \circ \widehat{f}$ belongs to $A(\widehat{G}, M_n)$ locally at every point of $\widehat{G} \cup \{\infty\}$. Fix $\zeta \in \widehat{G} \cup \{\infty\}$ and define $\widehat{f}(\infty) = 0$.

Regard $\mathcal{U}$ as a subset of $(M_n, \|\cdot\|_{hs}) = \mathbb{C}^{n^2}$ and let $\widehat{f}(\zeta) = w = (w_{11}, \ldots, w_{nn}) \in \mathbb{C}^{n^2}$. Choose $\varepsilon > 0$ such that F has the Taylor series expansion

$$F(z) = F(w) + \sum_\kappa A_\kappa (z - w)^\kappa$$

which converges absolutely for $\|z - w\|_{\mathbb{C}^{n^2}} < \varepsilon$, where $A_{(0,\ldots,0)} = 0$.

By Lemma 3.3.18, and by Lemma 3.3.19 if $\zeta = \infty$, we can find a function $g = (g_{ij}) \in L^1(G, M_n)$ with $\|g\|_1 < \varepsilon/n^2$ such that

$$\widehat{f}(\gamma) = \widehat{f}(\zeta) + \widehat{g}(\gamma)$$

in some neighbourhood of ζ. On G, consider the M_n-valued function

$$\sum_\kappa A_\kappa g^\kappa$$

where, for $\kappa = (\kappa_{11}, \ldots, \kappa_{nn})$, g^κ is the complex function

$$g^\kappa = g_{11}^{\kappa_{11}} * \cdots * g_{nn}^{\kappa_{nn}}$$

on G, and $g_{ij}^{\kappa_{ij}}$ is the κ_{ij}-times convolution $g_{ij} * \cdots * g_{ij}$. We have

$$\|g^\kappa\|_1 \leq \|g_{11}\|_1^{\kappa_{11}} \cdots \|g_{nn}\|_1^{\kappa_{nn}}$$

and $|g_{ij}(x)| \leq \|g(x)\|_{hs} \leq n\|g(x)\|_{M_n}$ implies

$$\left(\sum_{i,j} \|g_{ij}\|_1^2 \right)^{1/2} \leq \left(\sum_{i,j} n^2 \|g\|_1^2 \right)^{1/2} = (n^4 \|g\|_1^2)^{1/2} < \varepsilon.$$

Therefore the series

$$\sum_{\kappa} \|A_\kappa\| \|g_{11}\|_1^{\kappa_{11}} \cdots \|g_{nn}\|_1^{\kappa_{nn}}$$

converges. Hence the series $\sum_\kappa A_\kappa g^\kappa$ converges in $L^1(G, M_n)$ to a function, say, $h \in L^1(G, M_n)$. We have

$$\widehat{h} = \sum_\kappa A_\kappa \widehat{g_{11}}^{\kappa_{11}} \cdots \widehat{g_{nn}}^{\kappa_{nn}}$$

and

$$\begin{aligned}
F(\widehat{f}(\gamma)) &= F(\widehat{f}(\zeta)) + \sum_\kappa A_\kappa (\widehat{f}(\gamma) - \widehat{f}(\zeta))^\kappa \\
&= F(\widehat{f}(\zeta)) + \sum_\kappa A_\kappa \widehat{g}(\gamma)^\kappa \\
&= F(\widehat{f}(\zeta)) + \widehat{h}(\gamma)
\end{aligned}$$

in a neighbourhood of ζ and, we can find a function $\widehat{h}_1 \in A(\widehat{G}, M_n)$ which equals the constant $F(\widehat{f}(\zeta))$ in a smaller neighbourhood. This proves that $F \circ \widehat{f}$ belongs to $A(\widehat{G}, M_n)$ locally at every point of $\widehat{G} \cup \{\infty\}$.   □

We can now describe the L^p-spectrum in the abelian case.

Theorem 3.3.23. *Let G be abelian and let $\sigma \in M(G, M_n)$ be absolutely continuous. Then we have*

$$\mathrm{Spec}(T_\sigma, L^p(G, M_n)) = \overline{\Lambda\{\widehat{\sigma}(\pi) : \pi \in \widehat{G}\}}$$

for $1 \leq p \leq \infty$.

Proof. Let $\sigma = h \cdot \lambda$ for some $h \in L^1(G, M_n)$. We first consider $p = 1$. Suppose $\alpha \notin \overline{\Lambda\widehat{\sigma}(\widehat{G})} = \overline{\Lambda\widehat{h}(\widehat{G})}$. We show that

$$T_\sigma - \alpha I : L^1(G, M_n) \longrightarrow L^1(G, M_n)$$

is invertible. We have $\alpha \notin \Lambda(T_\sigma, L^1)$ for if $f \in L^1(G, M_n)$ satisfies $f * (\sigma - \alpha \delta_e) = 0$, then $\widehat{f}(\widehat{\sigma} - \alpha) = 0$ where, for all $\gamma \in \widehat{G}$, the matrix $\widehat{\sigma}(\gamma) - \alpha I$ is invertible in M_n since $\det(\widehat{\sigma}(\gamma) - \alpha I) \neq 0$ and hence $\widehat{f}(\gamma) = 0$.

It remains to show that $T_\sigma - \alpha$ is surjective. Consider the continuous function $\psi : M_n \longrightarrow \mathbb{C}$ given by

$$\psi(A) = \det(A - \alpha) \qquad (A \in M_n).$$

We note that the compact set $\psi\left(\overline{\widehat{h}(\widehat{G})}\right)$ does not contain 0. Otherwise, we have $0 = \psi(A)$ for some $A = \lim_\beta \widehat{h}(\gamma_\beta)$ with $\gamma_\beta \in \widehat{G}$. It follows that $\lim_\beta \det(\widehat{h}(\gamma_\beta) - \alpha) = \det(A - \alpha) = 0$ and one argues as in the proof of Lemma 3.3.10 to get a contradiction that $\alpha \in \overline{\Lambda\widehat{h}(\widehat{G})}$. We can therefore find an open set V in $\mathbb{C}$ containing $\psi\left(\overline{\widehat{h}(\widehat{G})}\right)$, but not 0. This gives an open set $\mathcal{U} = \psi^{-1}(V)$ in M_n containing $\overline{\widehat{h}(\widehat{G})}$ such that, for each $A \in \mathcal{U}$, the matrix $A - \alpha I$ is invertible in M_n.

We consider two cases : (i) $\alpha \neq 0$ and (ii) $\alpha = 0$.

Case (i). This occurs if $\widehat{G}$ is non-compact since $\widehat{h}$ vanishes at infinity and $0 \in \overline{\Lambda\widehat{h}(\widehat{G})}$ by Remark 3.3.11. We define a holomorphic map $F : \mathcal{U} \longrightarrow M_n$ by $F(z) = z(z-\alpha)^{-1}$. By Theorem 3.3.22, there exists $f \in L^1(G, M_n)$ such that $\widehat{f} = F(\widehat{h})$. Then we have $\widehat{f}(\gamma) = F(\widehat{h}(\gamma)) = \widehat{h}(\gamma)(\widehat{h}(\gamma) - \alpha)^{-1}$ for $\gamma \in \widehat{G}$.

Given $g \in L^1(G, M_n)$, we define

$$u = \frac{1}{\alpha}(g * f - g).$$

Then we have $(T_\sigma - \alpha I)(u) = g$ since

$$(u * (\sigma - \alpha \delta_e))\widehat{} = \widehat{u}(\widehat{h} - \alpha) = \frac{1}{\alpha}(\widehat{g}\widehat{f} - \widehat{g})(\widehat{h} - \alpha) = \frac{1}{\alpha}(\widehat{g}\widehat{f}(\widehat{h} - \alpha) - \widehat{g}\widehat{h} + \alpha\widehat{g}) = \widehat{g}.$$

Hence $T_\sigma - \alpha I$ is invertible and $\alpha \notin \mathrm{Spec}(T_\sigma, L^1)$. This proves $\mathrm{Spec}(T_\sigma, L^1) \subset \overline{\Lambda\widehat{\sigma}(\widehat{G})}$.

Case (ii). If $\alpha = 0$, then $\widehat{h}(\gamma)$ is invertible in M_n for all $\gamma \in \widehat{G}$ and we can apply Theorem 3.3.22 to the holomorphic function $F(z) = z^{-1}$ to obtain a function $f \in L^1(G, M_n)$ satisfying $\widehat{f}(\gamma) = \widehat{h}(\gamma)^{-1}$ for all $\gamma \in \widehat{G}$. Given $g \in L^1(G, M_n)$, we have $g = T_\sigma(u)$ for $u = g * f$. Hence T_σ is invertible which also proves $\mathrm{Spec}(T_\sigma, L^1) \subset \overline{\Lambda\widehat{\sigma}(\widehat{G})}$.

Now it follows from Proposition 3.3.17 that $\mathrm{Spec}(T_\sigma, L^\infty) = \mathrm{Spec}(T_\sigma, L^1) = \overline{\Lambda\widehat{\sigma}(\widehat{G})}$. The same conclusion for $1 < p < \infty$ follows from Lemma 3.3.2 and Proposition 3.3.17. $\qquad\square$

Remark 3.3.24. The above result extends the known description of the scalar L^p-spectrum for abelian groups: $\mathrm{Spec}(T_\sigma, L^p(G)) - \overline{\widehat{\sigma}(\widehat{G})}$ if σ is absolutely continuous. Without absolute continuity, the result is false for $p \neq 2$ (see, for example, [52, 66]).

Corollary 3.3.25. *Let G be a discrete abelian group and $\sigma \in M(G, M_n)$. Then we have* $\mathrm{Spec}(T_\sigma, L^\infty(G, M_n)) = \Lambda(T_\sigma, L^\infty(G, M_n))$.

Proof. Since $\widehat{G}$ is compact, the set $\Lambda\widehat{\sigma}(\widehat{G})$ is closed by continuity of $\widehat{\sigma}$ and the determinant function det. Hence the result follows from Theorem 3.3.23 and Proposition 3.3.16. $\qquad\square$

Given an abelian group G, the Fourier algebra $A(\widehat{G}) = (L^1(G))\widehat{}$ is abelian and its spectrum identifies with $\widehat{G}$. Hence the spectrum of each element $f \in A(\widehat{G})$ is the closure of $f(\widehat{G})$. The matrix Fourier algebra $A(\widehat{G}, M_n)$ is non-abelian and we have the following spectral result.

Corollary 3.3.26. *Let G be an abelian group. The quasi-spectrum of an element $\widehat{h}$ in the Banach algebra $A(\widehat{G}, M_n)$ is given by* $\mathrm{Spec}'\,\widehat{h} = \overline{\Lambda\widehat{h}(\widehat{G})}\cup\{0\}$.

Proof. First, assume that $\widehat{G}$ is non-compact. Then as remarked before, we have $0 \in \overline{\Lambda \widehat{h}(\widehat{G})}$. Let $\alpha \in \Lambda \widehat{h}(\widehat{G}) \backslash \{0\}$. We show $\widehat{h}/\alpha$ has no quasi-inverse in $A(\widehat{G}, M_n)$; otherwise, let $\widehat{g}$ be its quasi-inverse, then $\widehat{h}/\alpha + \widehat{g} = (\widehat{h}/\alpha)\widehat{g}$ implies

$$(\alpha I - \alpha \widehat{g}(\gamma))(\alpha I - \widehat{h}(\gamma)) = I \qquad (\gamma \in \widehat{G}).$$

Taking determinant both sides, we get a contradiction since $\det(\alpha I - \widehat{h}(\gamma_0)) = 0$ for some $\gamma_0 \in \widehat{G}$. Hence we have $\overline{\Lambda \widehat{h}(\widehat{G})} \subset \mathrm{Spec}' \widehat{h}$.

To see the reverse inclusion, we show that $\mathrm{Spec}' \widehat{h} \subset \mathrm{Spec}(T_\sigma, L^1(G, M_n))$ and invoke Theorem 3.3.23, where $\sigma = h \cdot \lambda$. Indeed, if $T_\sigma - \beta I : L^1(G, M_n) \longrightarrow L^1(G, M_n)$ is invertible for some $\beta \neq 0$, then we can find $f \in L^1(G, M_n)$ such that $f * h - \beta f = h$ giving $\widehat{f}(\widehat{h} - \beta) = \widehat{h}$ and $\widehat{h}/\beta + \widehat{f} = (\widehat{h}/\beta)\widehat{f}$. Hence $\widehat{h}/\beta$ has quasi-inverse $\widehat{f}$ and $\beta \notin \mathrm{Spec}' \widehat{h}$.

Finally, if $\widehat{G}$ is compact, then G is discrete and $A(\widehat{G}, M_n)$ is unital and the identity is the constant function on $\widehat{G}$ taking value $I \in M_n$. Similar arguments as above yield $\mathrm{Spec} \widehat{h} \subset \mathrm{Spec}(T_\sigma, L^1(G, M_n))$. This completes the proof. □

In connection with the results above, we consider, given $\sigma \in M(G, M_n)$ and $f \in L^1(G, M_n)$, the existence of solution to the matrix convolution equation $h * \sigma = f$ in $L^1(G, M_n)$.

Proposition 3.3.27. *Let G be abelian and $\sigma \in M(G, M_n)$. The following conditions are equivalent.*

(i) $T_\sigma : L^1(G, M_n) \longrightarrow L^1(G, M_n)$ *has dense range.*
(ii) $L_{\widetilde{\sigma}} : L^1(G, M_n) \longrightarrow L^1(G, M_n)$ *has dense range.*
(iii) $\det \widehat{\sigma}(\pi) \neq 0$ *for each $\pi \in \widehat{G}$.*

Proof. (i) $\Longrightarrow$ (iii). Condition (i) is equivalent to injectivity of $T_\sigma^* = L_{\widetilde{\sigma}} : L^\infty(G, M_n^*) \longrightarrow L^\infty(G, M_n^*)$ which, by Lemma 3.3.14, implies that $\det \widetilde{\widehat{\sigma}}(\pi) \neq 0$ for $\pi \in \widehat{G} \backslash \{\iota\}$, or, $\det \widehat{\sigma}(\pi) \neq 0$ for $\pi \in \widehat{G} \backslash \{\iota\}$. Pick $h \in L^1(G, M_n)$ such that

$$\det \widehat{h}(\iota) = \det \int_G h \, d\lambda \neq 0,$$

for instance, one can choose h to be a diagonal matrix with a diagonal entry $k \in L^1(G)$ satisfying $\int_G k \, d\lambda \neq 0$. Condition (i) gives a sequence (h_n) in $L^1(G, M_n)$ such that $(h_n * \sigma)$ is norm convergent to h. It follows that

$$\widehat{h}(\iota) = \lim_n \widehat{h_n * \sigma}(\iota) = \lim_n \widehat{h_n}(\iota)\widehat{\sigma}(\iota) \neq 0$$

which implies that $\det \widehat{\sigma}(\iota) \neq 0$.

(iii) $\Longrightarrow$ (i). We show that $L_{\widetilde{\sigma}}$ is injective. Let $f \in L^\infty(G, M_n^*)$ satisfy $\widetilde{\sigma} *_\ell f = 0$. Applying Lemma 3.3.14 to $\widetilde{\sigma}$, we conclude that f must take constant value $A \in M_n$, say. Then

$$\widehat{\sigma}(\iota)A = \widetilde{\sigma} *_\ell f = 0$$

implies that $A = 0$ since $\widehat{\sigma}(\iota)$ is invertible.

(ii) $\Longleftrightarrow$ (iii). Condition (ii) is equivalent to injectivity of the dual map $L_{\widetilde{\sigma}}^* = T_\sigma$: $L^\infty(G, M_n^*) \longrightarrow L^\infty(G, M_n^*)$, that is, $0 \notin \Lambda(T_\sigma, L^\infty(G, M_n^*)) = \Lambda(T_\sigma, L^\infty(G, M_n))$. The last equality holds because $M_n^* = (M_n, \|\cdot\|_1)$ and a bounded M_n^*-valued function can be regarded as a bounded M_n-valued function, and *vice versa*. $\qquad\square$

Corollary 3.3.28. *Let G be abelian and $\sigma \in M(G, M_n)$. The following conditions are equivalent.*

(i) $T_\sigma : L^1(G, M_n) \longrightarrow L^1(G, M_n)$ *is surjective.*
(ii) $L_{\widetilde{\sigma}} : L^1(G, M_n) \longrightarrow L^1(G, M_n)$ *is surjective.*
(iv) $0 \notin \operatorname{Spec} \sigma$.

Proof. By Lemma 3.3.2 and its proof, we have $\operatorname{Spec} \sigma = \operatorname{Spec}(L_{\widetilde{\sigma}}, L^1(G, M_n))$ and we only need to show (i) $\Longrightarrow$ (iv) $\Longleftarrow$ (ii).

(i) $\Longrightarrow$ (iv). By Proposition 3.3.27, we have $0 \notin \Lambda(T_\sigma, L^\infty(G, M_n))$ and hence $0 \notin \Lambda(T_\sigma, L^1(G, M_n))$ by Lemma 3.3.2. Therefore T_σ is invertible on $L^1(G, M_n)$.

Similar arguments yield (ii) $\Longrightarrow$ (iv). $\qquad\square$

Example 3.3.29. Let σ be the Gaussian measure on $\mathbb{R}$. Then the convolution operator $T_\sigma : L^1(\mathbb{R}, M_n) \longrightarrow L^1(\mathbb{R}, M_n)$ is not surjective, but has dense range. In fact, if G is a non-discrete abelian group and if $\sigma = f \cdot \lambda$ is absolutely continuous, then $T_\sigma : L^1(G, M_n) \longrightarrow L^1(G, M_n)$ is never surjective, for otherwise, we would have $f = h * \sigma = h * f$ for some $h \in L^1(G, M_n)$ which gives $\widehat{f}(\pi) = \widehat{h}(\pi)\widehat{f}(\pi)$, with $\widehat{f}(\pi) = \widehat{\sigma}(\pi)$ invertible for all $\pi \in \widehat{G}$ by Proposition 3.3.27, and hence $\widehat{h}(\pi) = I$ for all $\pi \in \widehat{G}$ which is impossible.

We note that surjectivity of $T_\sigma : L^1(G, M_n) \longrightarrow L^1(G, M_n)$ implies invertibility of $T_\sigma : L^p(G, M_n) \longrightarrow L^p(G, M_n)$, for $1 < p < \infty$, by Lemma 3.3.2 (iv); but the converse need not be true by Remark 3.3.24, since there are abelian groups G for which one can find $\mu \in M(G)$ with $\operatorname{Spec} \mu \setminus \operatorname{Spec}(T_\mu, L^2(G)) \neq \emptyset$.

We have the following criterion for surjectivity of T_σ on $L^2(G, M_n)$.

Lemma 3.3.30. *Let H be a Hilbert space and $T \in \mathcal{B}(H)$ with adjoint T^*. Then T is surjective if, and only if, $\operatorname{Spec} TT^* \subset [c, \infty)$ for some $c > 0$.*

Proof. We first note that the range $T(H)$ is closed if, and only if,

$$\operatorname{Spec} TT^* \subset \{0\} \cup [c, \infty)$$

for some $c > 0$ (cf. [6, p.95]).

Let T be surjective. Then $T^*(H)$ is closed by the above remark, and we only need to show $0 \notin \operatorname{Spec} TT^*$. Since $H = T^*(H) \oplus T^*(H)^\perp$, we have

$$H = TT^*(H) + T(T^*(H)^\perp) = TT^*(H).$$

By self-adjointness, TT^* is injective and $0 \notin \operatorname{Spec} TT^*$.

Conversely, $0 \notin \operatorname{Spec} TT^*$ implies that $H = TT^*(H)$ and T is surjective. $\qquad\square$

Corollary 3.3.31. *Let G be an abelian group and $\sigma \in M(G)$. The following conditions are equivalent.*

(i) $T_\sigma : L^2(G) \longrightarrow L^2(G)$ *is surjective.*

(ii) $|\widehat{\sigma}(\widehat{G})| \subset [c, \infty)$ *for some $c > 0$.*

Proof. Let $\overline{\sigma}$ be the complex conjugate of σ. We have $\operatorname{Spec} T_\sigma T_\sigma^* = \operatorname{Spec} T_\sigma T_{\overline{\sigma}} = \operatorname{Spec} T_{\widetilde{\overline{\sigma}} * \sigma} = \widehat{\widetilde{\overline{\sigma}} * \sigma}(\widehat{G})$, where $\widehat{\widetilde{\overline{\sigma}} * \sigma}(\widehat{G}) = |\widehat{\sigma}(\widehat{G})|^2$ since $\widetilde{\overline{\sigma}} * \sigma = \overline{\widehat{\sigma}}\widehat{\sigma} = |\widehat{\sigma}|^2$. Now Lemma 3.3.30 gives the result. □

For any group G, the operator $T_\sigma : L^1(G, M_n) \longrightarrow L^1(G, M_n)$ has dense range if, and only if, $0 \notin \Lambda(L_{\widetilde{\sigma}}, L^\infty(G, M_n^*)) = \Lambda(L_{\widetilde{\sigma}}, L^\infty(G, M_n)) = \Lambda(T_{\widetilde{\sigma}^T}, L^\infty(G, M_n))$, the latter equality follows from the fact that $(\widetilde{\sigma} *_\ell f)^T = f^T * \widetilde{\sigma}^T$ for every $f \in L^\infty(G, M_n)$.

Corollary 3.3.32. *Let $\sigma \in M(G, M_n)$ be such that $T_\sigma : L^1(G, M_n) \longrightarrow L^1(G, M_n)$ has dense range. Then 0 is not an eigenvalue of $\widehat{\widetilde{\sigma}}(\pi)$ for each $\pi \in \widehat{G}$.*

Proof. Injectivity of $L_{\widetilde{\sigma}} : L^\infty(G, M_n) \longrightarrow L^\infty(G, M_n)$ implies that 0 is not an eigenvalue of $\widehat{\widetilde{\sigma}}(\pi)$ for $\pi \in \widehat{G} \backslash \{\iota\}\}$, by Remark 3.3.15. The rest of the proof is similar to that of (i) $\Longrightarrow$ (iii) in Proposition 3.3.27. □

A complex-valued measure σ is called symmetric if $\widetilde{\sigma} = \sigma$. Extending this notion, we call an M_n-valued measure σ *symmetric* if $\widetilde{\sigma} = \sigma^T$. The above remarks give the following result.

Proposition 3.3.33. *Let $\sigma \in M(G, M_n)$ be symmetric. The following conditions are equivalent.*

(i) $T_\sigma : L^1(G, M_n) \longrightarrow L^1(G, M_n)$ *has dense range.*

(ii) $L_{\widetilde{\sigma}} : L^1(G, M_n) \longrightarrow L^1(G, M_n)$ *has dense range;*

(iii) $0 \notin \Lambda(T_\sigma, L^\infty(G, M_n))$.

Corollary 3.3.34. *Given a symmetric $\sigma \in M(G, M_n)$, the following conditions are equivalent.*

(i) $T_\sigma : L^1(G, M_n) \longrightarrow L^1(G, M_n)$ *is surjective.*

(ii) $0 \notin \operatorname{Spec} \sigma$.

Proof. Condition (i) implies $0 \notin \Lambda(T_\sigma, L^1(G, M_n))$. □

We now develop a device to study the spectrum of $T_\sigma : L^2(G, M_n) \longrightarrow L^2(G, M_n)$ for non-abelian groups G. We will show that T_σ^* identifies with an element of the C*-tensor product $C_r^*(G) \otimes M_{n^2}$ of the reduced group C*-algebra $C_r^*(G)$ and M_{n^2}. From this, we deduce several results about the spectrum $\operatorname{Spec}(T_\sigma, L^2(G, M_n))$. We first recall some basics of group C*-algebras. Since we use the *right* Haar measure λ on G, we define the group C*-algebra of G by the *right* regular representation $\rho : G \longrightarrow B(L^2(G))$ which is given by

$$\rho(x)h(y) = h(yx) \qquad (x, y \in G, h \in B(L^2(G))).$$

We note that $L^1(G, M_n)$ is a Banach $\star$-algebra in the convolution product and the involution

$$f^\star(x) = \triangle_G(x^{-1})f(x^{-1})^\star \qquad (x \in G)$$

where $\ast$ denotes the involution in M_n and $\triangle_G$ is the modular function of G. The regular representation ρ extends to a representation of the Banach algebra $M(G)$, still denoted by ρ:

$$\rho(\mu)(h) = \int_G \rho(x)h d\mu(x) = h \ast \tilde{\mu} \qquad (\mu \in M(G), h \in L^2(G))$$

and for $f \in L^1(G) \subset M(G)$, we have $\widehat{f\triangle_G} \in L^1(G)$ and

$$\rho(f)h(x) = \int_G f(y)h(xy)d\lambda(y) = h \ast \widehat{f\triangle_G}(x) \qquad (h \in L^2(G)).$$

The *reduced group C*-algebra* $C_r^*(G)$ is defined to be the norm closure $\overline{\rho(L^1(G))}$ in $B(L^2(G))$. The group C*-algebra $C^*(G)$ is the C*-completion of $L^1(G)$, that is, the completion with respect to the norm

$$\|f\|_c = \sup_\pi \{\|\pi(f)\|\}$$

where the supremum is taken over all $\star$-representations $\pi : L^1(G) \longrightarrow B(H_\pi)$. The regular representation ρ of $L^1(G)$ extends to a representation ρ' of $C^*(G)$ and we have $\rho'(C^*(G)) = C_r^*(G)$. Although ρ is injective on $L^1(G)$, its extension ρ' need not be so on $C^*(G)$. In fact, ρ' is faithful on $C^*(G)$ if and only if G is amenable.

To put our device for T_σ in perspective, we make use of C*-crossed products which extend the above construction of $C^*(G)$. Let $(\mathcal{A}, G, \beta)$ be a C*-dynamical system, that is, $\mathcal{A}$ is a C*-algebra, G a locally compact group and $\beta : t \in G \mapsto \beta_t \in$ Aut $\mathcal{A}$ is a homomorphism to the automorphism group Aut $\mathcal{A}$ of $\mathcal{A}$ such that the map $t \in G \mapsto \beta_t(x) \in \mathcal{A}$ is continuous for all $x \in \mathcal{A}$. We refer to [51] for an exposition of C*-dynamical systems and C*-crossed products.

Let $L^1(G, \mathcal{A})$ be the Lebesgue space of $\mathcal{A}$-valued λ-integrable functions on G. It is a Banach algebra with the following product and involution:

$$f \cdot h(x) = \int_G \beta_y(f(xy^{-1})h(y)d\lambda(y) \qquad (f, h \in L^1(G, \mathcal{A}), x \in G)$$

$$f^\star(x) = \triangle_G(x^{-1})\beta_x(f(x^{-1}))^\star \qquad (f \in L^1(G, \mathcal{A}), x \in G)$$

where $\ast$ denotes the involution in $\mathcal{A}$. The *C*-crossed product* $G \times_\beta \mathcal{A}$ is defined to be the C*-completion of $L^1(G, \mathcal{A})$. In particular, if $\mathcal{A} = \mathbb{C}$, then the action β is trivial, that is, each β_t is the identity map of $\mathcal{A}$, and the crossed product $G \times_\beta \mathbb{C}$ reduces to $C^*(G)$.

Given $\mathcal{A} \subset \mathcal{B}(H)$, let $L^2(G,H)$ be the Hilbert space of H-valued L^2 functions on G, with respect to the Haar measure λ. One can construct, as before, a *regular representation* $\beta' : G \times_\beta \mathcal{A} \longrightarrow \mathcal{B}(L^2(G,H))$ satisfying

$$\beta'(f)h(x) = \int_G \beta_x(f(y))(h(xy))d\lambda \tag{3.9}$$

for $f \in L^1(G,\mathcal{A}), h \in L^2(G,H)$ and $x \in G$. The image $\beta'(G \times_\beta \mathcal{A})$ is the *reduced C*-crossed product* $G \times_{\beta_r} \mathcal{A}$.

Given a C*-dynamical system $(\mathcal{A}, G, \iota)$ in which the action ι is trivial, the crossed product $G \times_\iota \mathcal{A}$ is the projective C*-tensor product $C^*(G) \otimes_{\max} \mathcal{A}$, while the reduced crossed product $G \times_{\iota r} \mathcal{A}$ is the injective C*-tensor product $C_r^*(G) \otimes_{\min} \mathcal{A}$.

We identify M_{n^2} with the unique C*-tensor product $M_n \otimes M_n$ and embed M_n into M_{n^2} via the map $A \in M_n \mapsto I_{M_n} \otimes A \in M_{n^2}$, where $I_{M_n} \otimes A : \mathbb{C}^{n^2} \longrightarrow \mathbb{C}^{n^2}$ is a linear map.

If we identify M_n with the vector space $\mathbb{C}^{n^2}$ via the map

$$(b_{ij}) \in M_n \mapsto (b_{11},\ldots,b_{n1},b_{12},\ldots,b_{n2},\ldots,b_{1n},\ldots,b_{nn}) \in \mathbb{C}^{n^2} \tag{3.10}$$

then for each $B = (b_{ij}) \in M_n = \mathbb{C}^{n^2}$ in (3.10), we have

$$(I_{M_n} \otimes A)(B) = \begin{pmatrix} A & & & \\ & A & & \\ & & \ddots & \\ & & & A \end{pmatrix} \begin{pmatrix} b_{11} \\ \vdots \\ b_{n1} \\ \vdots \\ b_{1n} \\ \vdots \\ b_{nn} \end{pmatrix} = AB \tag{3.11}$$

where AB is regarded as a vector in $\mathbb{C}^{n^2}$ as in (3.10).

Now, given an absolutely continuous $\sigma \in M(G, M_n)$, we are ready to identify the convolution operator $L_{\tilde{\sigma}} : L^2(G, M_n) \longrightarrow L^2(G, M_n)$ as an element in the reduced C*-crossed product $G \times_{\iota r} M_{n^2}$.

Proposition 3.3.35. *Let $\sigma \in M(G, M_n)$ be absolutely continuous. Let $M_{n,2}$ be the vector space M_n equipped with the Hilbert-Schmidt norm. Then the convolution operator $L_{\tilde{\sigma}} : L^2(G, M_{n,2}) \longrightarrow L^2(G, M_{n,2})$ is an element in the C*-tensor product $C_r^*(G) \otimes M_{n^2}$.*

Proof. Let $\sigma = f \cdot \lambda$ for some $f \in L^1(G, M_n)$. Let $\mathbf{1} \odot f : G \longrightarrow M_{n^2}$ be the function

$$(\mathbf{1} \odot f)(x) = I_{M_n} \otimes f(x) \qquad (x \in G).$$

Then we have $\mathbf{1} \odot f \in L^1(G, M_{n^2})$.

Consider the C*-dynamical system (M_{n^2}, G, ι) in which M_{n^2} acts on the Hilbert space $\mathbb{C}^{n^2}$ and the action ι is trivial. The regular representation

$$\iota' : G \times_\iota M_{n^2} \longrightarrow \mathcal{B}(L^2(G, \mathbb{C}^{n^2}))$$

is as defined in (3.9). We have, by (3.11),

$$
\begin{aligned}
\iota'(1 \odot f)h(x) &= \int_G (1 \odot f)(y)(h(xy))d\lambda(y) \qquad (h \in L^2(G, \mathbb{C}^{n^2}), x \in G) \\
&= \int_G (I_{M_n} \otimes f(y))(h(xy))d\lambda(y) \\
&= \int_G f(y)h(xy)d\lambda(y) \\
&= L_{\tilde{\sigma}}h(x)
\end{aligned}
$$

where $M_{n,2}$ is the Hilbert space $\mathbb{C}^{n^2}$. Hence $L_{\tilde{\sigma}} = \iota'(1 \odot f) \in \iota'(G \times_\iota M_{n^2}) = C_r^*(G) \otimes M_n$. □

Remark 3.3.36. In the above proof, if we regard $G \times_\iota M_{n^2}$ as the tensor product $C^*(G) \otimes M_{n^2}$, then the regular representation ι' is just the representation $\rho' \otimes 1$ of $C^*(G) \otimes M_{n^2}$, where ρ' is the regular representation of $C^*(G)$ and 1 the identity representation of M_{n^2}.

Proposition 3.3.35 provides us with a useful device to compute the spectrum $\operatorname{Spec}(T_\sigma, L^2(G, M_n))$. Indeed, we noted before that the spectrum does not change if one replaces the norm of M_n by the trace norm or the Hilbert-Schmidt norm, and therefore we have

$$
\begin{aligned}
&\operatorname{Spec}(T_\sigma, L^2(G, M_n)) \cup \{0\} \\
&= \operatorname{Spec}(T_\sigma^*, L^2(G, M_n^*)) \cup \{0\} \\
&= \operatorname{Spec}(L_{\tilde{\sigma}}, L^2(G, M_{n,2})) \cup \{0\} \\
&= \operatorname{Spec}_{\mathcal{B}(L^2(G, M_{n,2}))} L_{\tilde{\sigma}} \cup \{0\} \\
&= \operatorname{Spec}'_{\mathcal{B}(L^2(G, M_{n,2}))} L_{\tilde{\sigma}} \\
&= \operatorname{Spec}'_{C_r^*(G) \otimes M_{n^2}} L_{\tilde{\sigma}}
\end{aligned}
$$

where, we note that, given an element a in a C*-subalgebra $\mathcal{A}$ of another one $\mathcal{B}$, the two quasi-spectra $\operatorname{Spec}'_{\mathcal{A}} a$ and $\operatorname{Spec}'_{\mathcal{B}} a$ are equal. The C*-algebra $C_r^*(G) \otimes M_{n^2}$ has an identity if G is discrete, in which case, we have $\operatorname{Spec}(T_\sigma, L^2(G, M_n)) = \operatorname{Spec}_{C_r^*(G) \otimes M_{n^2}} L_{\tilde{\sigma}}$.

Corollary 3.3.37. *Let G be a discrete group such that $C_r^*(G)$ has no proper projection. Then the spectrum $\operatorname{Spec}(T_\sigma, L^2(G))$ is connected for each $\sigma \in M(G)$.*

Proof. Since $C_r^*(G)$ is projectionless, functional calculus implies that every element in $C_r^*(G)$, in particular $L_{\tilde{\sigma}}$, has a connected spectrum. □

It has been conjectured by Kadison that the reduced C*-algebra $C_r^*(G)$ of a torsion free discrete group G is projectionless. The free groups of at least two generators verify this conjecture, as well as some others (cf. [2, p.90]).

The *spectrum* of a C*-algebra A is defined to be the space $\widehat{A}$ of (equivalence classes) of non-zero irreducible representations $\pi : A \longrightarrow \mathcal{B}(H_\pi)$ of A [23, 3.1.5]. Given a self-adjoint element a in a unital C*-algebra A, we have

$$\operatorname{Spec}_A a = \bigcup_{\pi \in \widehat{A}} \operatorname{Spec}_{\mathcal{B}(H_\pi)} \pi(a)$$

(cf. [23, 3.3.5]). In fact, the above equality holds for an element $a \in A$ satisfying

$$\alpha \in \operatorname{Spec}_A a \Longleftrightarrow a - \alpha 1 \quad \text{has no left inverse in } A. \tag{3.12}$$

If A is non-unital, with unit extension $A_1 = A \oplus \mathbb{C}$, then we have the identification $\widehat{A}_1 = \widehat{A} \cup \{\omega\}$ where ω is the one-dimensional irreducible representation of A_1 annihilating A (cf. [23, 3.2.4]). In this case, for $a \in A$ satisfying (3.12) in A_1, we have the quasi-spectrum

$$\operatorname{Spec}_A' a = \operatorname{Spec}_{A_1} a = \bigcup_{\pi \in \widehat{A}_1} \operatorname{Spec} \pi(a)$$
$$= \bigcup_{\pi \in \widehat{A}} \operatorname{Spec} \pi(a) \cup \operatorname{Spec} \omega(a) = \bigcup_{\pi \in \widehat{A}} \operatorname{Spec} \pi(a) \cup \{0\}.$$

The spectrum $\widehat{C^*(G)}$ identifies with $\widehat{G}$ [23, 13.93] where each $\pi \in \widehat{G}$ is identified as the irreducible representation of $C^*(G)$ satisfying

$$\pi(f) = \int_G f(x)\pi(x)d\lambda(x) \qquad (f \in L^1(G) \subset C^*(G)).$$

The spectrum $\widehat{C_r^*(G)}$ identifies with the following closed subset of $\widehat{G}$, the *reduced dual* of G:

$$\widehat{G}_r = \{\tau\rho' : \tau \in \widehat{C_r^*(G)}\} = \{\pi \in \widehat{G} : \ker \pi \supset \ker \rho'\}$$

where ρ' is the right regular representation of $C^*(G)$. In general $\widehat{G}_r \neq \widehat{G}$, but they coincide if G is amenable [23, 18.3].

Lemma 3.3.38. *Let* $\sigma \in M(G, M_n)$ *be symmetric. Then for* $\alpha \in \mathbb{C}$, *we have* $\alpha \in \operatorname{Spec}(L_{\widetilde{\sigma}}, L^2(G, M_{n,2}))$ *if, and only if,* $L_{\widetilde{\sigma}} - \alpha I$ *has no left inverse in* $\mathcal{B}(L^2(G, M_{n,2}))$.

Proof. Let $\mu = \sigma - \alpha \delta_e$. Then $\widetilde{\mu} = \mu^T$ and $L_{\widetilde{\mu}} = L_{\widetilde{\sigma}} - \alpha I$.

It suffices to show that if $L_{\widetilde{\mu}}$ has a left inverse $S : L^2(G, M_{n,2}) \longrightarrow L^2(G, M_{n,2})$, then $L_{\widetilde{\mu}}$ has a right inverse. Taking dual, we have $I = L_{\widetilde{\mu}}^* S^* = T_\mu S^*$, that is, $f = S^* f * \mu$ for each $f \in L^2(G, M_{n,2})$. It follows that

$$f^T = (S^* f * \mu)^T = \mu^T *_\ell (S^* f)^T = \widetilde{\mu} *_\ell (S^* f)^T \qquad (f \in L^2(G, M_{n,2}))$$

by symmetry of μ. Define $R : L^2(G, M_{n,2}) \longrightarrow L^2(G, M_{n,2})$ by

$$R(f) = (S^* f^T)^T \qquad (f \in L^2(G, M_{n,2})).$$

Then we have

$$L_{\widetilde{\mu}} R(f) = \widetilde{\mu} *_\ell R(f) = \widetilde{\mu} *_\ell (S^* f^T)^T = f$$

for each $f \in L^2(G, M_{n,2})$. Hence R is a right inverse of $L_{\widetilde{\mu}}$. $\square$

Corollary 3.3.39. *Let G be a locally compact group and $\sigma \in M(G, M_n)$ be absolutely continuous and symmetric. Then we have*

$$\mathrm{Spec}\,(T_\sigma, L^2(G, M_n)) \cup \{0\} = \bigcup_{\pi \in \widehat{G}_r} \mathrm{Spec}\,\widehat{\sigma}(\pi) \cup \{0\}.$$

If G is discrete, $\{0\}$ can be removed.

Proof. By absolute continuity, we identify σ as a function in $L^1(G, M_n)$ and consider $L_{\widetilde{\sigma}}$ in $C_r^*(G) \otimes M_{n^2}$. By Lemma 3.3.38, $L_{\widetilde{\sigma}}$ satisfies (3.12) and hence

$$\mathrm{Spec}'_{C_r^*(G) \otimes M_{n^2}} L_{\widetilde{\sigma}} = \bigcup_{\gamma \in (C_r^*(G) \otimes M_{n^2})^\widehat{}} \mathrm{Spec}\,\gamma(L_{\widetilde{\sigma}}) \cup \{0\}$$

$$= \bigcup_{\tau \in \widehat{C_r^*(G)}} \mathrm{Spec}\,(\tau \otimes 1)(L_{\widetilde{\sigma}}) \cup \{0\}.$$

Let $\pi = \tau \rho' \in \widehat{G}_r \subset \widehat{G}$ where $\tau \in \widehat{C_r^*(G)}$ and $\rho' : C^*(G) \longrightarrow C_r^*(G)$ is the right regular representation.

Let $F \in L^1(G, M_{n^2})$ be the function

$$F = 1 \odot \sigma = \begin{pmatrix} \sigma & & \\ & \ddots & \\ & & \sigma \end{pmatrix}.$$

We can write $F = \sum_{ij} F_{ij} \otimes e_{ij}$, where $F_{ij} \in L^1(G)$ and $\{e_{ij}\}$ is the canonical matrix unit in M_{n^2}. By Remark 3.3.36 and as in the proof of Proposition 3.3.35, we have

$$(\tau \otimes 1)(L_{\widetilde{\sigma}}) = (\tau \otimes 1)(\iota'(1 \odot \sigma))$$

$$= (\tau \otimes 1)(\rho' \otimes 1)\left(\sum_{ij} F_{ij} \otimes e_{ij}\right)$$

$$= \sum_{ij} \tau \rho'(F_{ij}) \otimes e_{ij} = \sum_{ij} \pi(F_{ij}) \otimes e_{ij}$$

$$= \begin{pmatrix} \pi(\sigma) & & \\ & \ddots & \\ & & \pi(\sigma) \end{pmatrix} = I_{M_n} \otimes \pi(\sigma).$$

Hence $\mathrm{Spec}\,(\tau \otimes 1)(L_{\widetilde{\sigma}}) = \mathrm{Spec}\,(I_{M_n} \otimes \pi(\sigma)) = \mathrm{Spec}\,\pi(\sigma) = \mathrm{Spec}\,\widehat{\widetilde{\sigma}}(\pi)$, by (3.6).
It follows that

$$\mathrm{Spec}'_{C^*_r(G) \otimes M_{n^2}} L_{\widetilde{\sigma}} = \bigcup_{\pi \in \widehat{G}_r} \mathrm{Spec}\,\widehat{\widetilde{\sigma}} \cup \{0\}.$$

Hence, by (3.8), we have

$$\mathrm{Spec}\,(T_\sigma, L^2) \cup \{0\} = \mathrm{Spec}'_{C^*_r(G) \otimes M_{n^2}} L_{\widetilde{\sigma}}$$

$$= \bigcup_{\pi \in \widehat{G}_r} \mathrm{Spec}\,\widehat{\sigma^T}(\pi) \cup \{0\} = \bigcup_{\pi \in \widehat{G}_r} \mathrm{Spec}\,\widehat{\sigma}(\pi) \cup \{0\}.$$

If $C^*_r(G)$ has an identity, the above arguments can be applied to the spectrum of
$L_{\widetilde{\sigma}}$ instead of its quasi-spectrum. $\square$

Example 3.3.40. The Heisenberg group

$$\mathcal{H} = \left\{ \begin{pmatrix} 1 & x & z \\ 0 & 1 & y \\ 0 & 0 & 1 \end{pmatrix} : x, y, z \in \mathbb{R} \right\}$$

is amenable and we have (cf. [27, 6.51])

$$\widehat{\mathcal{H}} = \{\chi_{a,b} : a, b \in \mathbb{R}\} \cup \{\tau_t : t \in \mathbb{R} \setminus \{0\}\}$$

in which

$$\chi_{a,b} : (x, y, z) \in \mathcal{H} \mapsto e^{2\pi i(ax+by)} \in \mathbb{T},$$

$$\tau_t : \mathcal{H} \longrightarrow \mathcal{B}(L^2(\mathbb{R})) \quad \text{and} \quad \tau_t(x, y, z)f(w) = e^{2\pi i(tyw+tz)} f(w+x) \quad (f \in L^2(\mathbb{R}))$$

where an element in $\mathcal{H}$ is naturally denoted by (x, y, z).
    For $\sigma \in M(\mathcal{H})$, the set $\bigcup_{\tau \in \widehat{G}} \mathrm{Spec}\,\widehat{\sigma}(\tau)$ is given by

$$\left\{ \int_{\mathcal{H}} e^{-2\pi i(ax+by)} d\sigma(x, y, z) : a, b \in \mathbb{R} \right\} \bigcup \left\{ \mathrm{Spec} \int_{\mathcal{H}} \tau_t(-x, -y, xy-z) d\sigma(x, y, z) : t \neq 0 \right\}$$

which yields the spectrum $\mathrm{Spec}\,(T_\sigma, L^2(H))$ if σ is absolutely continuous and symmetric.
    If σ is the unit mass $\delta_{(x,0,z)}$ or $\delta_{(0,y,0)}$ where $y \neq 0$ and x or z is non-zero, then
the translation operator T_σ has spectrum $\mathrm{Spec}\,(T_\sigma, L^\infty(\mathcal{H})) = \mathbb{T}$. This follows from
Proposition 3.3.8 since $\bigcup_{\tau \in \widehat{G}} \mathrm{Spec}\,\widehat{\sigma}(\tau) = \mathbb{T}$ where

$$\widehat{\delta}_{(x,0,z)}(\chi_{a,b}) = e^{-2\pi iax}; \qquad \widehat{\delta}_{(0,y,0)}(\chi_{a,b}) = e^{-2\pi iby}$$

$$\widehat{\delta}_{(x,0,z)}(\tau_t)f(w) = e^{-2\pi itz} f(w-x); \qquad \widehat{\delta}_{(0,y,0)}(\tau_t)f(w) = e^{-2\pi ityw} f(w)$$

with $\mathrm{Spec}\,\widehat{\delta}_{(x,0,z)}(\tau_t) = e^{-2\pi itz}\widehat{\delta}_x(\mathbb{R})$ and $\mathrm{Spec}\,\widehat{\delta}_{(0,y,0)}(\tau_t) = \{e^{-2\pi ityw} : w \in \mathbb{R}\}$ (cf.
Example 2.1.13).

Example 3.3.41. An important problem in spectral geometry is the computation of the spectrum of the Laplacian. One can use Corollary 3.3.39 to compute the spectrum of a discrete Laplacian $\mathcal{L}_d$ of a homogeneous graph.

A graph (V, E) is called a *homogeneous graph* [19] if the vertex set V is a homogeneous space of a discrete group G with a graph condition, by which we mean G acts transitively on V by a right action $(v, g) \in V \times G \mapsto vg \in V$ so that V is represented as a right coset space G/H of G by a finite subgroup H and the edge set E is described by a finite subset $K = K^{-1} \subset G$ in that $(v, u) \in E$ if, and only if, $u = va$ for some $a \in K$. We denote a homogeneous graph by (V, K), with the edge generating set K, and by $|K|$ the cardinality of K. We note that (V, K) is a Cayley graph if H reduces to the identity of G.

Given a homogeneous graph (V, K) with weight given by a positive symmetric measure σ on G supported by K, satisfying $|K| = \sum_{a \in K} \sigma\{a\}$, the discrete Laplacian $\mathcal{L}_d$, acting on real or complex functions f on V, is defined by

$$\mathcal{L}_d f(v) = \frac{1}{|K|} \sum_{a \in K} (f(v) - f(va))\sigma\{a\}$$

$$= f * \left(\delta_e - \frac{\sigma}{|K|} \right)(v) \qquad (v \in V)$$

which is a convolution operator on $V = G/H$, as defined in (3.15). The operator $I - \mathcal{L}_d$ is the *transition operator*. For a Cayley graph (V, K), it follows from Corollary 3.3.39 that the L^2-spectrum of $\mathcal{L}_d$ is given by

$$1 - \bigcup \left\{ \mathrm{Spec} \left(\sum_{a \in K} \sigma\{a\}|K|^{-1}\pi(a) \right) : \pi \in \widehat{G} \right\}.$$

If we let $\mathcal{L}_d$ act on M_n-valued functions on V, the L^2-spectrum of $\mathcal{L}_d$ is an example of a *vibrational spectrum* [20] for the graph (V, K) and in this case, one can also describe the spectrum using Corollary 3.3.39.

Recently, Corollary 3.3.39 has been extended to the setting of homogeneous spaces in [7, Theorem 2.3] which can be used to describe the spectrum of $\mathcal{L}_d$ for any homogeneous graph (V, K).

It may be of interest to note that the L^2-spectrum of certain transition operator on the discrete lamplighter group, the wreath product $\mathbb{Z}_2 \mathrm{wr} \mathbb{Z}$, has been used in [35] to construct a counterexample to a conjecture of Atiyah concerning the L^2-Betti numbers of closed manifolds.

Corollary 3.3.42. *Let G be a finite group and let $\sigma \in M(G, M_n)$ be symmetric. We have*

$$\Lambda(T_\sigma, \ell^p(G, M_n)) = \mathrm{Spec}\,(T_\sigma, \ell^p(G, M_n)) = \mathrm{Spec}\,(T_\sigma, \ell^2(G, M_n)) = \Lambda\{\widehat{\sigma}(\pi) : \pi \in G\}$$

for all $p \in [0, 1]$

Proof. Since G is finite, each $\pi \in \widehat{G}_r = \widehat{G}$ and $\ell^p(G, M_n)$ are finite-dimensional and hence Corollary 3.3.39 yields the result. $\qquad\square$

The symmetry condition can be removed from the above result. In fact, the result is true for compact groups and absolutely continuous σ.

Proposition 3.3.43. *Let G be a compact group and $\sigma \in M(G, M_n)$. Then we have*

$$\mathrm{Spec}\,(T_\sigma, L^p(G, M_n)) \supset \bigcup_{\pi \in \widehat{G}} \mathrm{Spec}\,\widehat{\sigma}(\pi) \supset \Lambda(T_\sigma, L^p(G, M_n))$$

for all $p \in [1, \infty]$.

Proof. The last inclusion follows from Lemma 3.3.2 (i) and Proposition 3.3.8. For the first inclusion, we need only consider $1 < p < \infty$ by Proposition 3.3.8. Since G is compact, we have $L^p(G, M_n) \subset L^1(G, M_n)$ and all irreducible representations of G are finite dimensional. Let $\alpha \in \mathrm{Spec}\,\widehat{\sigma}(\pi)$ for some $\pi \in \widehat{G}$, where $\mathrm{Spec}\,\widehat{\sigma}(\pi) = \mathrm{Spec}\,\widehat{\sigma^T}(\pi)$ by (3.8). Then $\widehat{\sigma^T}(\pi)$ is a matrix and we have $\det(\widehat{\sigma^T}(\pi) - \alpha I_{M_n \otimes \mathcal{B}(H_\pi)}) = 0$. Since $\det \pi(e) = 1$, we can find a compact neighbourhood K of e such that $\det \pi(x^{-1}) > 1/2$ for all $x \in K$.

Suppose, for contradiction, that $L_{\sigma^T} - \alpha I : L^p(G, M_n) \longrightarrow L^p(G, M_n)$ is invertible. Then there exists $h \in L^p(G, M_n)$ such that

$$\sigma^T *_\ell h - \alpha h = \begin{pmatrix} \chi_K & & \\ & \ddots & \\ & & \chi_K \end{pmatrix}$$

where χ_K is the characteristic function of K.

Since $h \in L^1(G, M_n)$, we have

$$\widehat{\sigma^T}\widehat{h} - \alpha\widehat{h} = \begin{pmatrix} \widehat{\chi_K} & & \\ & \ddots & \\ & & \widehat{\chi_K} \end{pmatrix}$$

on $\widehat{G}$. In particular, we have

$$\widehat{\sigma^T}(\pi)\widehat{h}(\pi) - \alpha\widehat{h}(\pi) = \begin{pmatrix} \int_K \pi(x^{-1})d\lambda(x) & & \\ & \ddots & \\ & & \int_K \pi(x^{-1})d\lambda(x) \end{pmatrix}$$

which gives the contradiction

$$0 = \det(\widehat{\sigma^T}(\pi) - \alpha)\det\widehat{h}(\pi) = \det((\widehat{\sigma^T}(\pi) - \alpha)\widehat{h}(\pi))$$

$$= \left(\int_K \det\pi(x^{-1})d\lambda(x) \right)^n > \frac{1}{2^n}\lambda(K)^n > 0.$$

This proves non-invertibility of $L_{\sigma^T} - \alpha I$, that is, $\alpha \in \mathrm{Spec}\,(L_{\sigma^T}, L^p) = \mathrm{Spec}\,(T_\sigma, L^p)$, by Lemma 3.3.7. $\qquad\qquad\qquad\qquad\qquad\qquad\qquad\qquad\qquad\qquad\qquad\qquad\qquad\qquad\qquad\square$

Corollary 3.3.44. *Let G be a compact group and $\sigma \in M(G, M_n)$. If σ is absolutely continuous, then we have, for all $p \in [0, \infty]$,*

$$\mathrm{Spec}\,(T_\sigma, L^p(G, M_n)) = \overline{\Lambda(T_\sigma, L^p(G, M_n))} = \overline{\bigcup_{\pi \in \widehat{G}} \mathrm{Spec}\,\widehat{\sigma}(\pi)}.$$

Proof. This follows from the fact that T_σ is a compact operator in this case, by the proof of Theorem 3.2.1. $\qquad\qquad\qquad\qquad\qquad\qquad\qquad\qquad\qquad\qquad\qquad\qquad\qquad\square$

Without absolute continuity of σ, the above result is false, as noted in Remark 3.3.24.

In the remaining section, we study the eigenspaces of the convolution operator $T_\sigma : L^p(G, M_n) \longrightarrow L^p(G, M_n)$ for $\sigma \in M(G, M_n)$.

Definition 3.3.45. Let $1 \leq p \leq \infty$ and $\sigma \in M(G, M_n)$. For each $\alpha \in \mathbb{C}$, we define the space

$$H_\alpha(T_\sigma, L^p(G, M_n)) = \{ f \in L^p(G, M_n) : f * \sigma = \alpha f \}$$

which may be written as $H_\alpha(T_\sigma, L^p)$ for short. If $\alpha \in \Lambda(T_\sigma, L^p(G, M_n))$, that is, if α is an eigenvalue of T_σ, then $H_\alpha(T_\sigma, L^p(G, M_n))$ is the α-eigenspace of T_σ in $L^p(G, M_n)$.

By abuse of language, we call $H_\alpha(T_\sigma, L^p)$ the 'α-eigenspace' of T_σ even if $\alpha \notin \Lambda(T_\sigma, L^p(G, M_n))$. The α-eigenspace $H_\alpha(L_\sigma, L^p(G, M_n))$ of $L_\sigma : L^p(G, M_n) \longrightarrow L^p(G, M_n)$ is defined likewise.

Plainly $H_\alpha(T_\sigma, L^p)$ is a left invariant subspace of $L^p(G, M_n)$. One is interested in the dimension and the description of $H_\alpha(T_\sigma, L^p)$. For compact groups and absolutely continuous σ, the α-eigenspaces are finite dimensional for $\alpha \neq 0$. We discuss later spectral synthesis for eigenfunctions for abelian groups.

Proposition 3.3.46. *Let G be a compact group and $\sigma \in M(G, M_n)$ be absolutely continuous. Then T_σ is a compact operator on $L^p(G, M_n)$ for each $p \in [1, \infty]$ and hence $\dim H_\alpha(T_\sigma, L^p(G, M_n)) < \infty$ for all $\alpha \neq 0$.*

Proof. This follows from Theorem 3.2.1. $\qquad\qquad\qquad\qquad\qquad\qquad\qquad\qquad\qquad\qquad\qquad\qquad\square$

Example 3.3.47. For the probability measure $d\sigma(x) = \dfrac{\sin^2 x}{\pi x^2} dx$ on $\mathbb{R}$, the eigenspace $H_0(T_\sigma, L^2(\mathbb{R}))$ is infinite dimensional. We have the Fourier transform

$$\widehat{\sigma}(t) = \frac{2 - |t|}{2} \chi_{[-2,2]}(t) \qquad (t \in \mathbb{R})$$

and $\mathrm{Spec}(T_\sigma, L^2(\mathbb{R})) = \widehat{\sigma}(\mathbb{R}) = [0, 1]$. By [10, Corollary 3.14], $1 \notin \Lambda(T_\sigma, L^2(\mathbb{R}))$, but $0 \in \Lambda(T_\sigma, L^2(\mathbb{R}))$. Indeed, for $a < b$, let $g_{a,b} \in L^2(\mathbb{R})$ be defined by

$$g_{a,b}(x) = \frac{e^{-iax} - e^{-ibx}}{ix}$$

whose Fourier transform $\widehat{g_{a,b}}$ equals $2\pi\chi_{(a,b)}$ on $\mathbb{R}\backslash\{a,b\}$ and it follows that

$$\widehat{g_{a,b} * \sigma} = \widehat{g_{a,b}}\,\widehat{\sigma} = 0$$

if $(a,b) \cap [-2,2] = \emptyset$ and the eigenspace $H_0(T_\sigma, L^2(\mathbb{R}))$ contains

$$\{g_{a,b} : (a,b) \cap [-2,2] = \emptyset\}$$

which is infinite dimensional.

Definition 3.3.48. An eigenspace $H_\alpha(T_\sigma, L^p(G, M_n))$ is said to be *trivial* if it consists of only constant functions.

We note that, if $\sigma(G) \neq I_{M_n}$, then the constant function $\mathbf{1} : G \longrightarrow M_n$ taking value I_{M_n} does not belong to $H_1(T_\sigma, L^\infty(G, M_n))$. Nevertheless, $H_1(T_\sigma, L^\infty(G, M_n))$ can still contain a non-zero constant function $f(\cdot) = A \in M_n$ where, for instance, $I_{M_n} - A$ can be taken to be the support projection of $I_{M_n} - \sigma(G)$ (cf. [10, Example 1]).

Definition 3.3.49. For $\alpha = \|\sigma\|$, the functions in $H_\alpha(T_\sigma, L^p(G, M_n))$ are called the M_n-valued L^p *σ-harmonic* functions on G.

By normalizing, it suffices to study the space $H_1(T_\sigma, L^p(G, M_n))$ of σ-harmonic functions for $\|\sigma\| = 1$. Thus, in this case, $1 \notin \Lambda(T_\sigma, L^p(G, M_n))$, equivalently, $H_1(T_\sigma, L^p(G, M_n)) = \{0\}$, denotes the absence of a non-zero M_n-valued L^p σ-harmonic function on G. If σ is an adapted probability measure on G, then we have $1 \notin \Lambda(T_\sigma, L^p(G))$ for $p < \infty$ unless G is compact, as shown in [10].

Example 3.3.50. Let $a \in \mathbb{R}\backslash\{0\}$ and consider the *non-adapted* probability measure $\sigma = \frac{1}{2}(\delta_a + \delta_{-a})$ on $\mathbb{R}$. We have

$$\Lambda(T_\sigma, L^\infty) = \widehat{\sigma}(\widehat{\mathbb{R}}) = \{\cos ax : x \in \mathbb{R}\} = [-1, 1]$$

and the 1-eigenspace $H_1(T_\sigma, L^\infty(\mathbb{R}))$ is infinite dimensional [13, Example 2.7.3].

For $\|\sigma\| = 1$, the triviality of $H_1(T_\sigma, L^\infty(G, M_n))$ is a *Liouville type* theorem for bounded harmonic functions on G. It has been shown in [16] that, given a nilpotent group G, if $\sigma \in M(G, M_n)$ is positive, non-degenerate and $\|\sigma\| = 1$, then $H_1(T_\sigma, L^\infty(G, M_n))$ is trivial (see also [40]). A Liouville theorem has also been proved for almost connected [IN]-groups in [15]. For arbitrary p, we have the following result.

Proposition 3.3.51. *Let G be a compact group and let $\sigma \in M(G, M_n)$ be positive, adapted and $\|\sigma\| = 1$. Then $H_1(T_\sigma, L^p(G, M_n))$ is trivial for $1 \leq p \leq \infty$.*

Proof. Since G is compact, we have $L^p(G,M_n) \subset L^1(G,M_n)$ and it suffices to consider the case for $p = 1$. Let $f \in H_1(T_\sigma, L^1(G,M_n))$. Let $\{u_\beta\}_\beta$ be a bounded approximate identity in $L^1(G)$ and let

$$\psi_\beta = \begin{pmatrix} u_\beta & & \\ & \ddots & \\ & & u_\beta \end{pmatrix}.$$

Then $\psi_\beta * f \longrightarrow f$ in $L^1(G,M_n)$. Each $\psi_\beta * f$ is a bounded continuous M_n-valued σ-harmonic function on G and must be constant, by [9, Proposition 21]. It follows that f is constant. $\qquad\qquad\square$

Example 3.3.52. In contrast to the scalar case where $H_1(T_\sigma, L^\infty(G)) \supset \mathbb{C}1$ for a probability measure σ, one can have $H_1(T_\sigma, L^\infty(G,M_n)) = \{0\}$ for a positive matrix measure σ with $\|\sigma\| = 1$. Let F_2 be the free group on two generators a and b. Let $\sigma \in M(G,M_2)$ be supported on $\{a,b\}$ and defined by

$$\sigma\{a\} = \begin{pmatrix} \frac{1}{2} & 0 \\ 0 & 0 \end{pmatrix}, \quad \sigma\{b\} = \begin{pmatrix} 0 & 0 \\ 0 & \frac{1}{2} \end{pmatrix}.$$

Then σ is a positive adapted measure on F_2 with $\|\sigma\| = 1$. Given $f = (f_{ij}) \in H_\sigma^\infty(G,M_2)$, we have $f_{i1} * \sigma_{11} = f_{i1}$ and $f_{i2} * \sigma_{22} = f_{i2}$, where $\|f_{ij} * \sigma_{jj}\| \leq \|f_{ij}\|\|\sigma_{jj}\|$ and $\|\sigma_{11}\| = \|\sigma_{22}\| < 1$ imply $f_{ij} = 0$ for all i,j. Hence $H_\sigma^\infty(G,M_2) = \{0\}$.

We now consider the question of synthesis for eigenfunctions in the case of abelian groups. For each $f \in L^p(G,M_n)$, we denote by

$$\ell(f) = \text{lin}\{\ell_x f : x \in G\} \subset L^p(G,M_n)$$

the linear span of the left translations of f in $L^p(G,M_n)$.

We first observe that $\overline{\ell(f)} \neq L^p(G,M_n)$ for each eigenfunction $f \in H_\alpha(T_\sigma, L^p)$. Indeed, if $\overline{\ell(f)} = L^p(G,M_n)$ for some 0-eigenfunction f, then for each $h \in C_c(G,M_n)$, there is a sequence (f_n) in $\ell(f)$ converging to h in $L^p(G,M_n)$ which gives $h * \sigma = \lim_n f_n * \sigma = 0$ and

$$\int_G h\, d\widetilde{\sigma} = h * \sigma(e) = 0,$$

contradicting $\sigma \neq 0$. Since α is an eigenvalue of T_σ if, and only if, 0 is an eigenvalue of $T_{\sigma - \alpha\delta_e}$, the assertion is true for any α-eigenfunction f.

Given $f \in L^p(G,M_n)$ and $\varphi \in L^q(G,M_n)$, we have $f * \widetilde{\varphi}(x) = (\ell_{x^{-1}} f) * \widetilde{\varphi}(e)$ for each $x \in G$. Hence $\overline{\ell(f)} = L^p(G,M_n)$ if, and only if, $\varphi = 0$ for any $\varphi \in L^q(G,M_n)$ satisfying $f * \widetilde{\varphi} = 0$.

It follows that, for each eigenfunction $f \in H_\alpha(T_\sigma, L^p(G,M_n))$, there exists a nonzero function $\varphi \in L^q(G,M_n)$ such that $f * \widetilde{\varphi} = 0$. If G is abelian, then Wiener's Tauberian theorem implies that for each eigenfunction $f \in H_\alpha(T_\sigma, L^1(G))$, its Fourier transform $\widehat{f}$ has a zero in $\widehat{G}$.

To highlight the idea behind synthesis, we consider the scalar case. First, for $\sigma \in M(\mathbb{R})$ with compact support, the continuous 0-eigenfunctions in $L^p(\mathbb{R})$ are the mean periodic functions on $\mathbb{R}$ which have been analysed completely in the classic paper [58] of Schwartz. In particular, these functions can be synthesized from the so-called exponential polynomials, in other words, in the space $C(\mathbb{R})$ of complex continuous functions on $\mathbb{R}$, the subspace of mean periodic functions is the closed linear span of the mean periodic exponential polynomials. This result has been extended to locally compact abelian groups by Gilbert [31] and Elliot [26]. We now apply these results to our eigenspaces.

A complex function on $\mathbb{R}$ is called an *exponential polynomial* if it is of the form

$$p(x)e^{i\gamma x}$$

where $p(x)$ is a polynomial with complex coefficients and $\gamma \in \mathbb{C}$. More generally, a complex function on $\mathbb{R}^n$ is called an *exponential polynomial* if it is of the form

$$p(x_1,\ldots,x_n)\exp(i\gamma_1 x_1 + \cdots + i\gamma_n x_n)$$
$$= \sum_{0 \leq i_1,\ldots,i_n \leq k} a_{i_1 \cdots i_n} x_1^{i_1} \cdots x_n^{i_n} \exp(i\gamma_1 x_1 + \cdots + i\gamma_n x_n).$$

It is clear that a bounded exponential polynomial on $\mathbb{R}^n$ reduces to a constant multiple of a character.

Proposition 3.3.53. *Let $\sigma \in M(\mathbb{R})$ have compact support. For $\alpha \in \Lambda(T_\sigma, L^\infty(\mathbb{R}))$ and $f \in H_\alpha(T_\sigma, L^\infty(\mathbb{R})) \cap C(\mathbb{R})$, we have*

$$\overline{\ell(f)} = \overline{\mathrm{lin}}\{\widehat{\mathbb{R}} \cap \overline{\ell(f)}\}$$

and hence,

$$H_\alpha(T_\sigma, L^\infty) = \overline{\mathrm{lin}}\{\widehat{\mathbb{R}} \cap H_\alpha(T_\sigma, L^\infty)\}$$

where the closure is the weak* closure.

Proof. Let $\mu = \sigma - \alpha \delta_e$. Then μ has compact support in $\mathbb{R}$ and $H_\alpha(T_\sigma, L^\infty) = H_0(T_\mu, L^\infty)$.

Let $f \in H_0(T_\mu, L^\infty) \cap C(\mathbb{R})$ so that $f * \mu = 0$. By [58], there is a net (p_β) in $C(\mathbb{R})$ converging to f uniformly on compact subsets of $\mathbb{R}$, where each p_β is a linear combination of exponential polynomials $p_{\beta_1}, \ldots, p_{\beta_k}$ which belong to the closure of $\ell(f)$, in the topology of $C(\mathbb{R})$. We have $p_{\beta_j} * \mu = 0$ for $j = 1, \ldots, k$. Since $\ell(f) \subset L^\infty(\mathbb{R})$, each p_{β_j} must be bounded and is therefore a constant multiple of a character. Also p_{β_j} belongs to the weak* closure $\overline{\ell(f)}$ for if it is the limit of a net (h_γ) in $\ell(f)$ in the topology of $C(\mathbb{R})$, then for each $k \in C_c(\mathbb{R})$, we have

$$\langle k, h_\gamma \rangle - \langle k, p_{\beta_j} \rangle = \int_{\mathrm{supp}\, k} k(x)(h_\gamma(x) - p_{\beta_j}(x))d\lambda(x) \longrightarrow 0.$$

Hence each p_β belongs to $\mathrm{lin}\{\widehat{G} \cap \overline{\ell(f)}\}$ which implies $f \in \overline{\mathrm{lin}}\{\widehat{G} \cap \overline{\ell(f)}\}$ and the result follows. □

The above result depends on the property of spectral synthesis in $\mathbb{R}$, proved in [58], that f can be approximated by exponential polynomials belonging to the closure of $\ell(f)$ in $C(\mathbb{R})$. This property is lost in $\mathbb{R}^n$ for $n > 1$. For instance, if $f : \mathbb{R}^2 \longrightarrow \mathbb{R}$ is the function $f(x,y) = x + y$, then the closure of $\ell(f)$ in $C(\mathbb{R}^2)$ contains no exponential polynomials except the constants. Nevertheless, it is still true that each $f \in C(\mathbb{R}^n)$ satisfying $f * \sigma = \alpha f$ can be approximated by exponential polynomials ψ satisfying $\psi * \sigma = \alpha \psi$ although ψ need not lie in the closure of $\ell(f)$. In fact, this is even true for locally compact abelian groups, due to the results of [26, 31]. To explain the details, we first describe the exponential polynomials on an abelian group G. A *real character* on G is a continuous homomorphism from G to the additive group $\mathbb{R}$. A complex function on G is called an *exponential polynomial* if it is of the form

$$p(\chi_1(x),\ldots,\chi_j(x))\tau(x) \qquad (x \in G)$$

where $p(\cdot)$ is polynomial with a finite number of variables and complex coefficients, $\chi_1,\ldots,\chi_j$ are real characters on G, and τ is a generalized character on G.

We note that a non-zero real character χ on G must be unbounded, for if $\chi(x) \neq 0$, then $|\chi(x^n)| = n|\chi(x)| \to \infty$ as $n \to \infty$. If $p(\chi_1,\ldots,\chi_j)\tau$ is bounded, then it must be a constant multiple of a character on G. Indeed, since $\tau(\cdot) \neq 0$, the function $p(\chi_1,\ldots,\chi_j)$ must be bounded which implies that the product $\chi_1 \cdots \chi_j = 0$ for otherwise, there exists $y \in G$ such that $\chi_1(y) \cdots \chi_j(y) \neq 0$, giving

$$|\chi_1(y^n)^{i_1} \cdots \chi_j(y^n)^{i_j}| = n^{i_1 + \cdots + i_j}|\chi_1(y)^{i_1} \cdots \chi_j(y)^{i_j}| \to \infty \qquad (n \to \infty)$$

if $i_1 + \cdots + i_j \neq 0$. Hence p reduces to a constant and the boundedness of τ implies that τ must be a character.

Let $P(G)$ be the set of exponential polynomials on an abelian group G.

Proposition 3.3.54. *Let G be an abelian group and let $\sigma \in M(G)$ have compact support. For each $\alpha \in \Lambda(T_\sigma, L^p(G))$ where $1 \leq p \leq \infty$, we have*

$$H_\alpha(T_\sigma, L^p) \cap C(G) \subset \overline{\mathrm{lin}}^c\{\psi \in P(G) : \psi * \sigma = \alpha\psi\}$$

where '$-^c$' denotes the closure in $C(G)$.

Proof. As before, the measure $\mu = \sigma - \delta_e$ has compact support and given $f \in H_\alpha(T_\sigma, L^p) \cap C(G)$, we have $f * \mu = 0$. By [31, Theorem 3.2], there is a net (ψ_β) in $C(G)$ converging to f uniformly on compact subsets of G, and each ψ_β is a linear combination of exponential polynomials $\psi \in P(G)$ satisfying $\psi * \mu = 0$. Hence we have

$$\psi_\beta \in \mathrm{lin}\{\psi \in P(G) : \psi * \sigma = \alpha\psi\}$$

which completes the proof. □

We note that $H_\alpha(T_\sigma, L^\infty) \cap C(G) = H_\alpha(T_\sigma, L^\infty)$ if σ is absolutely continuous.

For $1 \le p \le \infty$, we denote as usual by $L_{loc}^p(G)$ the space of Borel functions f on G satisfying $f|_K \in L^p(K)$ for every compact subset $K \subset G$. The topology of $L_{loc}^p(G)$ is defined by the norms $\| \cdot \|_{L^p(K)}$ from compact subsets K of G. We have $L^p(G) \subset L_{loc}^p(G)$.

Corollary 3.3.55. *Let G be an abelian group and let $\sigma \in M(G)$ have compact support. For each $\alpha \in \Lambda(T_\sigma, L^p(G))$ where $1 \le p < \infty$, we have*

$$H_\alpha(T_\sigma, L^p) \subset \overline{\lin}^{loc} \{ \psi \in P(G) : \psi * \sigma = \alpha\psi \}$$

where '$—^{loc}$' denotes the closure in $L_{loc}^p(G)$.

Proof. Let $f \in H_\alpha(T_\sigma, L^p)$ and let $\varepsilon > 0$. Choose $u \in C_c(G)$ such that $\| u * f - f \|_p < \varepsilon$. We have $u * f \in H_\alpha(T_\sigma, L^p) \cap C(G)$ and by Proposition 3.3.54, there exists a net (ψ_β) in $\lin \{ \psi \in P(G) : \psi * \sigma = \alpha\psi \}$, converging to $u * f$ uniformly on compact sets in G. Given a compact subset K of G, there exists β_0 such that $\| \psi_\beta - u * f \|_{L^p(K)} < \varepsilon$ for $\beta \ge \beta_0$ and hence

$$\| \psi_\beta - f \|_{L^p(K)} \le \| \psi_\beta - u * f \|_{L^p(K)} + \| u * f - f \|_{L^p(K)} < 2\varepsilon$$

for $\beta \ge \beta_0$. $\qquad\qquad\qquad\qquad\qquad\qquad\qquad\qquad\qquad\qquad\qquad\square$

We now consider L^p harmonic functions for arbitrary groups G. It has been shown in [9, Proposition 14] that, for $\|\sigma\| = 1$, the eigenspace $H_1(T_\sigma, L^\infty(G, M_n))$ is the range of a contractive projection $P : L^\infty(G, M_n) \longrightarrow L^\infty(G, M_n)$. There are two interesting consequences of this result. First, the space $H_1(T_\sigma, L^\infty(G, M_n))$ carries a Jordan algebraic structure which will be discussed further in Section 3.4. The second consequence is that, if G is non-amenable, then $H_1(T_\sigma, L^\infty(G, M_n)) \ne M_n \mathbf{1}$, for positive σ with $\|\sigma\| = 1$ [9, Corollary 19]. The existence of a contractive projection P onto the 1-eigenspace is also true for $1 < p < \infty$, shown in the following result which extends [10, Theorem 2.3], with analogous proof. We outline the main steps of the arguments.

Proposition 3.3.56. *Let $\sigma \in M(G, M_n)$ with $\|\sigma\| = 1$ and let $1 < p < \infty$. Then there is a contractive projection $P : L^p(G, M_n) \longrightarrow L^p(G, M_n)$ with range $H_1(T_\sigma, L^p(G, M_n))$ and P commutes with left translations. Further, the projection P is the dual map of a contractive projection $Q : L^q(G, M_n) \longrightarrow L^q(G, M_n)$ and $H_1(T_\sigma, L^p(G, M_n)) = H_1(L_{\widetilde{\sigma}}, L^q(G, M_n))^*$.*

Proof. The convolution operator $T_\sigma : L^p(G, M_n) \longrightarrow L^p(G, M_n)$ is weakly continuous when $L^p(G, M_n)$ is equipped with the weak topology. For $n = 1, 2, \ldots$, we have

$$T_{\sigma^n} = \overbrace{T_\sigma \circ \cdots \circ T_\sigma}^{n-times}.$$

Let $\mathcal{K} = \overline{co}\{T_{\sigma^n} : n = 1, 2, \ldots\}$ be the closed convex hull of $\{T_{\sigma^n} : n = 1, 2, \ldots\}$ with respect to the product topology of $L^p(G, M_n)^{L^p(G, M_n)}$ where $L^p(G, M_n)$ is equipped with the weak topology. Then $\mathcal{K}$ is compact. Define $\Phi : \mathcal{K} \longrightarrow \mathcal{K}$ by

$$\Phi(\Lambda)(f) = \Lambda(f) * \sigma \qquad (\Lambda \in \mathcal{K}, f \in L^p(G,M_n)).$$

It is straightforward to verify that Φ is well-defined and continuous. Therefore, by the Schauder-Tychonoff fixed-point theorem (cf. [24, p. 456]), there exists $P \in \mathcal{K}$ such that $\Phi(P) = P$ which is the required contractive projection. Since T_{σ^n} commutes with left translations, so does P.

We note that $P(f * \sigma) = P(f) * \sigma = P(f)$ for each $f \in L^p(G,M_n)$ since $T_{\sigma^n}(f * \sigma) = T_{\sigma^n}(f) * \sigma$.

Next, apply the same construction as above to the left convolution operator

$$L_{\widetilde{\sigma}} : f \in L^q(G,M_n) \mapsto \widetilde{\sigma} *_{\ell} f \in L^q(G,M_n)$$

to yield a contractive projection

$$Q : L^q(G,M_n) \longrightarrow L^q(G,M_n)$$

with range $H_1(L_{\widetilde{\sigma}}, L^q(G,M_n))$, and satisfying $Q(\widetilde{\sigma} *_{\ell} g) = \widetilde{\sigma} *_{\ell} Q(g) = Q(g)$ for $g \in L^q(G,M_n)$. We show that $P = Q^*$. Let $f \in L^p(G,M_n)$. Then for each $g \in L^q(G,M_n)$, we have

$$\langle g, Q^* f * \sigma \rangle = \langle \widetilde{\sigma} *_{\ell} g, Q^* f \rangle = \langle Q(\widetilde{\sigma} *_{\ell} g), f \rangle$$
$$= \langle Qg, f \rangle = \langle g, Q^* f \rangle.$$

Hence $Q^* f * \sigma = Q^* f$. Likewise one can show $\widetilde{\sigma} *_{\ell} P^* g = P^* g$ for each $g \in L^q(G,M_n)$. We now have $PQ^* = P$ since

$$\langle g, PQ^* f \rangle = \langle QP^* g, f \rangle = \langle P^* g, f \rangle = \langle g, Pf \rangle$$

for $g \in L^q(G,M_n)$ and $f \in L^p(G,M_n)$. Therefore $Pf = PQ^* f = Q^* f$.

Finally, as in the proof of [10, Corollary 2.5], we have $H_1(L_{\widetilde{\sigma}}, L^q(G,M_n))^* = Q(L^q(G,M_n))^* \simeq L^p(G,M_n)/Q(L^q(G,M_n))^{\perp} \simeq H_1(T_{\sigma}, L^p(G,M_n))$. $\qquad \square$

Remark 3.3.57. By the above construction of P, there is a net of measures (μ_{α}) in the convex hull of $\{\sigma^n : n = 1, 2, \ldots\}$ such that

$$P(f) = \text{w}^*\text{-}\lim_{\alpha} f * \mu_{\alpha}$$

for every $f \in L^p(G,M_n)$. We note that P could be 0 by Example 3.3.52.

The above construction does not apply to the case $p = 1$. Nevertheless, we will prove a dimension result for the eigenspace $H_1(T_{\sigma}, L^1(G,M_n))$. We need to prove the following lemma first.

Lemma 3.3.58. *Let $\sigma \in M(G,M_n)$ be a positive, adapted measure with $\|\sigma\| = 1$. Then for every $\pi \in \widehat{G}\setminus\{\imath\}$, the operator $I - \pi(\sigma)$ is invertible in $M_n \otimes \mathcal{B}(H_{\pi})$, where I is the identity operator.*

Proof. For compact groups G, this result was proved in [9, Lemma 20]. We only need to remove the compactness assumption in [9, Lemma 20] which was used to ensure $\dim \pi < \infty$. Fix $\pi \in \widehat{G} \backslash \{\iota\}$. It suffices to show that $I - \pi(\sigma)$ has a left inverse in $M_n \otimes \mathcal{B}(H_\pi)$. Indeed, since $(1 \otimes \pi)(\sigma \otimes 1) = (\sigma \otimes 1)(1 \otimes \pi)$ and $\sigma(\cdot)^* = \sigma(\cdot)$, the same arguments would imply that

$$I - \pi(\sigma)^* = I - \int_G (1 \otimes \pi)(x^{-1}) d(\sigma \otimes 1)(x)$$

has a left inverse, that is, $I - \pi(\sigma)$ has a right inverse.

Now, if $I - \pi(\sigma)$ has no left inverse, then $(M_n \otimes \mathcal{B}(H_\pi))(I - \pi(\sigma))$ is a proper weakly closed left ideal of $M_n \otimes \mathcal{B}(H_\pi)$ and hence there is a proper projection $p \in M_n \otimes \mathcal{B}(H_\pi)$ such that $(M_n \otimes \mathcal{B}(H_\pi))(I - \pi(\sigma)) = (M_n \otimes \mathcal{B}(H_\pi))p$. Therefore $I - \pi(\sigma) = (I - \pi(\sigma))p$ which gives $\pi(\sigma)(I - p) = I - p$ and we can pick a unit vector $\xi \in (I - p)(\mathbb{C}^n \otimes H_\pi)$. It follows that $\pi(\sigma)\xi = \xi$ and now, analogous to the proof of [9, Lemma 20], one obtains

$$\int_G \langle (1 \otimes \pi)(x)\xi, \xi \rangle d|\sigma|(x) = 1$$

which implies $\operatorname{Re}\langle (1 \otimes \pi)(x)\xi, \xi \rangle = 1$ for all $x \in \operatorname{supp}|\sigma|$ and hence $(1 \otimes \pi)(x)\xi = \xi$ for all $x \in G$, by adaptedness of $|\sigma|$.

Let $\{e_1, \dots, e_n\}$ be the standard basis of $\mathbb{C}^n$. Then $\xi = \sum_k e_k \otimes \xi_k$ for some $\xi_1, \dots, \xi_n \in H_\pi$ and we have

$$\sum_k e_k \otimes \xi_k = \sum_k (1 \otimes \pi)(x)(e_k \otimes \xi_k) = \sum_k e_k \otimes \pi(x)\xi_k$$

which implies $\pi(x)\xi_k = \xi_k$ for all $x \in G$, and in particular, for some $\xi_k \neq 0$. Hence $\pi = \iota$ by irreducibility of π, contradicting $\pi \in \widehat{G} \backslash \{\iota\}$. This completes the proof. $\qquad \square$

The following result generalizes [10, Theorem 3.12].

Proposition 3.3.59. *Let $\sigma \in M(G, M_n)$ be a positive, adapted measure with $\|\sigma\| = 1$. Then $\dim H_1(T_\sigma, L^1(G, M_n)) \leq n^2$ and n^2 is the best possible bound.*

Proof. Let $f \in H^1_\sigma(G, M_n)$. Then, for all $\pi \in \widehat{G} \backslash \{\iota\}$, we have $\pi(f) = \pi(f * \sigma) = \pi(f)\pi(\sigma)$ and hence $\pi(f)(I - \pi(\sigma)) = 0$ which implies $\pi(f) = 0$ by Lemma 3.3.58. Let

$$L^1_0(G, M_n) = \left\{ h \in L^1(G, M_n) : \int_G h \, d\lambda = 0 \right\}$$

which is a closed subspace of $L^1(G, M_n)$. Note that $\widehat{f}(\iota) = \int_G f \, d\lambda$ and the above yields

$$H^1_\sigma(G, M_n) \cap L^1_0(G, M_n) = \{0\}.$$

Pick $h \in L^1(G)$ such that $\int_G h \, d\lambda \neq 0$. Then

$$\begin{pmatrix} h & & \\ & \ddots & \\ & & h \end{pmatrix} \notin L_0^1(G, M_n).$$

For any $g = (g_{ij}) \in L^1(G, M_n)$, let

$$a_{ij} = \left(\int_G g_{ij} d\lambda \right) \left(\int_G h d\lambda \right)^{-1}.$$

Then

$$g + L_0^1(G, M_n) = (a_{ij}) \begin{pmatrix} h & & \\ & \ddots & \\ & & h \end{pmatrix} + L_0^1(G, M_n).$$

Hence $L_0^1(G, M_n)$ has co-dimension n^2 in $L^1(G, M_n)$ and $\dim H_1(T_\sigma, L^1(G, M_n)) \leq n^2$.

If G is compact and σ is diagonal with each diagonal entry the same probability measure on G, then $f = (f_{ij}) \in H_1(T_\sigma, L^1(G, M_n))$ if, and only if, each f_{ij} is constant. Therefore we have $\dim H_1(T_\sigma, L^1(G, M_n)) = n^2$ in this case. □

Example 3.3.60. Let μ be an adapted probability measure on a locally compact group $G \neq \{e\}$ and let $\sigma \in M(G, M_2)$ be given by

$$\sigma = \begin{pmatrix} \mu & 0 \\ 0 & \delta_e \end{pmatrix}.$$

Then σ is an adapted positive M_2-valued measure on G, but $\|\sigma\| = 2$. We have $\dim H_1(T_\sigma, L^1(G, M_2)) = \infty$, indeed, it contains the functions

$$\begin{pmatrix} 0 & f \\ 0 & h \end{pmatrix}$$

for every $f, h \in L^1(G)$.

Example 3.3.61. Let $\sigma = \delta_i$ be the unit mass at $i = \sqrt{-1}$ in the circle group $\mathbb{T}$. Then σ is not adapted and a continuous function f is in $H_1(T_\sigma, L^p(\mathbb{T}))$ if, and only if, $f(z) = f(-iz)$. Hence $H_1(T_\sigma, L^p(\mathbb{T}))$ is infinite dimensional as it contains the functions $\{z^{4n} : n = 1, 2, \ldots\}$.

We have the following characterisation of harmonic functions for nilpotent groups.

Proposition 3.3.62. *Let G be a nilpotent group and let $\sigma \in M(G, M_n)$ be positive, symmetric and $\|\sigma\| = 1$. Then we have, for $1 \leq p \leq \infty$,*

$$H_1(T_\sigma, L^p(G, M_n)) = \{f \in L^p(G, M_n) : f_{a^{-1}} = f = f(\cdot)\sigma(G), \forall a \in \operatorname{supp} \sigma\}.$$

Proof. Given $f \in L^p(G, M_n)$ satisfying the condition on the right-hand side of the above equality, we have

$$\int_G f(xy^{-1})d\sigma(y) = \int_{\text{supp}\,\sigma} f(x)d\sigma(y) = f(x)\sigma(G) = f(x).$$

To show the reverse inclusion, let G_σ be the closed subgroup of G generated by supp $|\sigma|$. Since σ is symmetric, we have $|\widetilde{\sigma}| = |\sigma^T| = |\sigma|$ and $(\text{supp } |\sigma|)^{-1} = \text{supp } |\sigma|$ and hence $|\sigma|$ is a non-degenerate measure on the nilpotent group G_σ. Each bounded left uniformly continuous M_n-valued σ-harmonic function f on G restricts to a σ-harmonic function on G_σ, and by [16, Theorem 4], f is constant on G_σ. In particular, we have

$$f(a^{-1}) = f(e) \qquad (a \in \text{supp}\,\sigma).$$

Now pick any $f \in H_1(T_\sigma, L^p(G, M_n))$. For each $\psi \in L^q(G, M_n)$, the function $\widetilde{\psi} * f$ is bounded and left uniformly continuous, by Lemma 3.1.1, and also it is σ-harmonic. Hence we have, for each $a \in \text{supp}\,\sigma$,

$$\langle f - f_{a^{-1}}, \psi \rangle = \text{Tr}\,(\widetilde{\psi} * f)(e) - \text{Tr}\,(\widetilde{\psi} * f)(a^{-1}) = 0$$

which yields $f - f_{a^{-1}} = 0$ in $L^p(G, M_n)$. This, together with the equation $f = f * \sigma$, then implies $f(\cdot) = f(\cdot)\sigma(G)$.   □

Under the conditions of the above result, $H_1(T_\sigma, L^\infty(G, M_n))$ is a subalgebra of $L^\infty(G, M_n)$ which is not always true in general. We study the non-associative algebraic structures of the eigenspace $H_1(T_\sigma, L^\infty(G, M_n))$ in the next section.

3.4 Jordan Structures in Harmonic Functions

Let $\sigma \in M(G, M_n)$ with $\|\sigma\| = 1$. In this section, we study the Jordan structure in the space $H_1(T_\sigma, L^\infty(G, M_n)) = \{f \in L^\infty(G, M_n) : f * \sigma = f\}$ of bounded M_n-valued σ-harmonic functions on G, and discuss applications to harmonic functions on Riemannian symmetric spaces. It has been shown in [9] that the existence of a contractive projection from $L^\infty(G, M_n)$ onto its subspace $H_1(T_\sigma, L^\infty(G, M_n))$ induces a Jordan ternary algebraic structure on $H_1(T_\sigma, L^\infty(G, M_n))$ which is non-associative and is usually different from that of $L^\infty(G, M_n)$. It is natural to ask when these two structures coincide, that is, when $H_1(T_\sigma, L^\infty(G, M_n))$ is a subalgebra or a Jordan sub-triple of $L^\infty(G, M_n)$. We will consider the scalar case of $H_1(T_\sigma, L^\infty(G))$ for a *complex* measure σ. The matrix space $H_1(T_\sigma, L^\infty(G, M_n))$, but with a *positive* measure σ, has been studied in [14].

It would be useful to explain first the background of Jordan structures for motivation. The close relationship between Jordan algebras and differential geometry is well-known [43], in particular, Jordan structures occur naturally in the study of Riemannian symmetric spaces. It is therefore interesting that Jordan structures also occur in the 1-eigenspace $H_1(T_\sigma, L^\infty(G, M_n))$ of the convolution operator T_σ on

$L^\infty(G,M_n)$, which is closely related to harmonic functions on Riemannian symmetric spaces, as we will explain below.

Let Ω be a simply connected Riemannian symmetric space. Then Ω is a product

$$\Omega = \Omega_0 \times \Omega_+ \times \Omega_-$$

where Ω_0 is Euclidean, Ω_+ is of compact type and Ω_- is of non-compact type [37, p.244]. Since Ω_0 and Ω_+ have non-negative sectional curvatures, the bounded harmonic functions on these manifolds are constant by a result of Yau [65], and from the viewpoint of harmonic functions, we will only be concerned with symmetric spaces of non-compact type.

We now explain how Jordan structures arise in symmetric spaces. Recall that a Riemannian symmetric space is a connected Riemannian manifold M in which every point x is an isolated fixed point of an involutive isometry $s_x : M \longrightarrow M$ (which is necessarily unique). Let Ω be a Riemannian symmetric space. Then it is diffeomorphic, and hence identified with, the right coset space G/K of a Lie group G by a maximal compact subgroup K, where G is the identity component of the isometry group of Ω and K is the isotropy subgroup $\{g \in G : gx = x\}$ at a point $x \in \Omega$. Let $\mathfrak{g}$ be the Lie algebra of G and let $Ad : G \longrightarrow Aut(\mathfrak{g})$ be the adjoint map. Then the adjoint $Ad(s_x) : \mathfrak{g} \longrightarrow \mathfrak{g}$ is an involutive automorphism, giving a Cartan decomposition $\mathfrak{g} = \mathfrak{k} \oplus \mathfrak{p}$ where

$$\mathfrak{k} = \{X \in \mathfrak{g} : Ad(s_x)(X) = X\}$$

and

$$\mathfrak{p} = \{X \in \mathfrak{g} : Ad(s_x)(X) = -X\}.$$

Moreover, $\mathfrak{p}$ identifies with the tangent space $T_x\Omega$ at $x \in \Omega$.

If Ω is non-compact and is Hermitian, then in the above construction, the tangent space $\mathfrak{p} = T_x\Omega$ has the structure of a *Jordan triple system* and Ω identifies with a convex domain in $\mathfrak{p}$ via the Harish-Chandra realization (cf. [47, 57]). In fact, this construction can even be extended to infinite dimensional manifolds in which case $\mathfrak{p}$ becomes a JB*-*triple* in a suitably chosen norm and Ω identifies with the open unit ball of $\mathfrak{p}$. We refer to [41, 63] for details and to [11] for some recent related results. For our purpose, we only need to explain the concept of a JB*-triple.

A *JB*-triple* is a complex Banach space Z, equipped with a *Jordan triple product* $\{\cdot,\cdot,\cdot\} : Z \times Z \times Z \longrightarrow Z$ which is symmetric and linear in the outer variables, conjugate linear in the middle variable and satisfies the Jordan triple identity

$$\{a,b,\{x,y,z\}\} = \{\{a,b,x\},y,z\} - \{x,\{b,a,y\},z\} + \{x,y,\{a,b,z\}\},$$

and for each $v \in Z$, the linear map

$$D(v,v) : z \in Z \longmapsto \{v,v,z\} \in Z$$

is Hermitian, that is, $\|e^{itD(v,v)}\| = 1$ for all $t \in \mathbb{R}$, and has non-negative spectrum with $\|D(v,v)\| = \|v\|^2$.

Example 3.4.1. The upper half-plane $\{z \in \mathbb{C} : \operatorname{Im} z > 0\}$, with the hyperbolic metric, is a symmetric space diffeomorphic to the coset space $SL(2,\mathbb{R})/SO(2)$ and is of non-compact type. The Lie algebra $\mathfrak{g}$ of $SL(2,\mathbb{R})$ is the algebra of 2×2 real matrices with trace 0, and in the Cartan decomposition $\mathfrak{g} = \mathfrak{k} \oplus \mathfrak{p}$, the subalgebra $\mathfrak{k}$ consists of skew-symmetric matrices while the subspace $\mathfrak{p}$ consists of symmetric matrices, which has a complex structure

$$J : \begin{pmatrix} y & x \\ x & -y \end{pmatrix} \mapsto \begin{pmatrix} x & -y \\ -y & -x \end{pmatrix}$$

and is a JB*-triple with the Jordan triple product

$$\{X,Y,Z\} = \frac{1}{2}(XYZ + ZYX) \qquad (X,Y,Z \in \mathfrak{p}).$$

To complete the picture, we note that a bounded domain in a complex Banach space is symmetric if, and only if, it is biholomorphic to the open unit ball of a JB*-triple [41]. We refer to [63] for further details of JB*-triples and symmetric Banach manifolds.

The space $L^\infty(G, M_n)$ is a JB*-triple with the Jordan triple product

$$\{f,g,h\} = \frac{1}{2}(fg^*h + hg^*f)$$

where g^* denotes the usual involution $*$ in $L^\infty(G, M_n)$:

$$g^*(x) := g(x)^* \in M_n \qquad (x \in G).$$

In fact, any C*-algebra $\mathcal{A}$ is a JB*-triple with the following triple product:

$$\{a,b,c\} = \frac{1}{2}(ab^*c + cb^*a) \qquad (a,b,c \in \mathcal{A}).$$

The following result has been proved in [9].

Lemma 3.4.2. *Let G be a locally compact group and let $\sigma \in M(G, M_n)$ with $\|\sigma\| = 1$. Then the 1-eigenspace $H_1(T_\sigma, L^\infty(G, M_n))$ of T_σ on $L^\infty(G, M_n)$ is a JB*-triple. If σ is a probability measure on G, then $H_1(T_\sigma, L^\infty(G))$ is an abelian C*-algebra.*

Proof. We describe the Jordan triple product in $H_1(T_\sigma, L^\infty(G, M_n))$, but refer to [9] for details. By [9, Proposition 14], there is a contractive projection $P : L^\infty(G, M_n) \longrightarrow H_1(T_\sigma, L^\infty(G, M_n))$ which induces the JB*-triple product

$$\{f,g,h\} = \frac{1}{2}P(fg^*h + hg^*f)$$

on $H_1(T_\sigma, L^\infty(G, M_n))$, using [28].

If σ is a probability measure, then the constant function $\mathbf{1}$ is in $H_1(T_\sigma, L^\infty(G))$ which becomes a unital abelian C*-algebra in the product

$$f \cdot h := \{f, \mathbf{1}, h\}.$$

$\square$

Given a probability measure σ on G, it has been shown in [13, Theorem 2.2.17] that the C*-product $f \cdot h$ in $H_1(T_\sigma, L^\infty(G))$, for uniformly continuous f and h, is given by

$$(f \cdot h)(x) = \lim_\alpha \int_G f(xy^{-1}) h(xy^{-1}) d\mu_\alpha(y) \qquad (x \in G)$$

where (μ_α) is a net in the convex hull of $\{\sigma^n : n = 1, 2, \ldots\}$. We now show that, modulo a projection, the C*-product is pointwise.

Proposition 3.4.3. *Let σ be a probability measure on G. Then there is a projection $z \in C_b(G)^{**}$ such that*

$$(f \cdot h)(x) z(\varepsilon_x) = f(x) h(x) z(\varepsilon_x) \qquad (x \in G)$$

for $f, h \in H_1(T_\sigma, L^\infty(G)) \cap C(G)$, where $\varepsilon_x \in C_b(G)^*$ is the evaluation map at $x \in G$.

Proof. We observe that $H_1(T_\sigma, L^\infty(G)) \cap C(G)$ is an abelian C*-algebra in the product $f \cdot h$. Therefore the identity map $\iota : H_1(T_\sigma, L^\infty(G)) \cap C(G) \longrightarrow C_b(G)$ is a linear isometry between C*-algebras, and by [18, Proposition 2.2; Theorem 3.10], there is a projection $z \in C_b(G)^{**}$ such that

$$\iota(f \cdot h) z = \iota(f) \iota(h) z \qquad (f, h \in H_1(T_\sigma, L^\infty(G)) \cap C(G))$$

where the product on the right hand side is that of $C_b(G)^{**}$. $\square$

A JB*-triple is called a JBW*-*triple* if it has a predual. Since T_σ is weak* continuous on $L^\infty(G, M_n)$, it follows that $H_1(T_\sigma, L^\infty)$ is weak* closed in $L^\infty(G, M_n)$ and is a JBW*-triple.

Now we consider harmonic functions on a symmetric space $\Omega = G/K$ of noncompact type. Let Δ be a G-invariant second order elliptic differential operator on Ω, vanishing on constants. Such an operator is called a *Laplace operator* in [29]. Furstenberg [29] has shown that there is a K-invariant absolutely continuous probability measure σ on G such that a bounded continuous function f on Ω satisfies $\Delta f = 0$ if, and only if,

$$f(Ka) = \int_G f(Kya) d\sigma(y) \qquad (Ka \in \Omega = G/K) \qquad (3.13)$$

where K-*invariance* of σ means $d\sigma(kx) = d\sigma(xk) = d\sigma(x)$ for $k \in K$ (see also [30, Theorem 5]). Let $q : G \longrightarrow G/K$ be the quotient map. Then (3.13) can be written as

$$\widetilde{f \circ q} = \widetilde{f \circ q} * \sigma \qquad (3.14)$$

where we recall that $\widetilde{f \circ q}$ denotes the function $\widetilde{f \circ q}(x) = f \circ q(x^{-1})$. As K is compact, we assume that the Haar measure λ on G is chosen so that $\lambda(K) = 1$ and the Haar measure on K is the restriction of λ to K. Also, there is a G-invariant measure

v on G/K, unique up to a constant multiple and can be chosen so that $v = \lambda \circ q^{-1}$ (cf. [27, p.57]). In fact, v is a Riemannian measure on Ω. Let $L^p(\Omega)$ be the Lebesgue spaces of v, for $1 \le p \le \infty$. Since Ω is complete, all $L^p(\Omega)$ Δ-harmonic functions on Ω are constant, for $1 < p < \infty$, by a result of Yau [65]. We will discuss the case for $p = 1, \infty$ below.

The above discussion leads naturally first to the consideration of the homogeneous space of an *arbitrary* locally compact group G by a compact subgroup K. In this case, we fix a G-invariant measure $v = \lambda \circ q^{-1}$ on $\Omega = G/K$ as before and let $L^p(\Omega, M_n)$ be the Lebesgue spaces, with respect to v, of M_n-valued L^p functions on Ω, for $1 \le p \le \infty$. Then the map

$$j : f \in L^p(\Omega, M_n) \longmapsto f \circ q \in L^p(G, M_n)$$

is a well-defined isometric embedding by the change-of-variable formula

$$\int_\Omega \|f(\omega)\|^p dv(\omega) = \int_G \|f \circ q(x)\|^p d\lambda(x).$$

We define a linear map $Q : L^p(G, M_n) \longrightarrow L^p(\Omega, M_n)$ by

$$Q(g)(Kx) = \int_K g(yx)d\lambda(y) \qquad (g \in L^p(G, M_n)).$$

Then Q is a contraction because Jensen's inequality gives

$$\|Q(g)\|_p^p = \int_\Omega \|Q(g)(Kx)\|_{M_n}^p dv(Kx)$$

$$= \int_G \left\| \int_K g(yx)d\lambda(y) \right\|_{M_n}^p d\lambda(x)$$

$$\le \int_G \int_K \|g(yx)\|_{M_n}^p d\lambda(y)d\lambda(x)$$

$$= \int_K \int_G \|g(x)\|_{M_n}^p \Delta_G(y^{-1})d\lambda(x)d\lambda(y) = \|g\|_p^p.$$

Further, Q is surjective since one verifies readily that Qj is the identity map on $L^p(\Omega, M_n)$. It follows that

$$jQ : L^p(G, M_n) \longrightarrow L^p(G, M_n)$$

is a contractive projection with range $j(L^p(\Omega, M_n))$.

Let $\sigma \in M(G, M_n)$. Then the convolution operator $T_\sigma : L^p(G, M_n) \longrightarrow L^p(G, M_n)$ induces the operator $T'_\sigma = QT_\sigma j$ on $L^p(\Omega, M_n)$ in the following diagram:

$$L^p(G,M_n) \xrightarrow{T_\sigma} L^p(G,M_n)$$

$$j\uparrow \qquad\qquad \downarrow Q$$

$$L^p(\Omega,M_n) \xrightarrow{T_\sigma'} L^p(\Omega,M_n).$$

Actually, we have

$$T_\sigma'(f)(Kx) = \int_G f(Kxy^{-1})d\sigma(y) \tag{3.15}$$

and we call T_σ' the *convolution operator* on $L^p(\Omega,M_n)$ defined by $\sigma \in M(G,M_n)$. The above integral is denoted by $f * \sigma(Kx)$. By Fubini theorem, we also have $T_\sigma'Q = QT_\sigma$ and $jT_\sigma' = T_\sigma j$ which will be used repeatedly in the proof below.

Lemma 3.4.4. *Let $\sigma \in M(G,M_n)$ and $1 \le p \le \infty$. Then we have*

(i) $\mathrm{Spec}(T_\sigma',L^p(\Omega,M_n)) \subset \mathrm{Spec}(T_\sigma,L^p(G,M_n))$;
(ii) $\Lambda(T_\sigma',L^p(\Omega,M_n)) \subset \Lambda(T_\sigma,L^p(G,M_n))$;
(iii) $Q(H_\alpha(T_\sigma,L^p(G,M_n))) = H_\alpha(T_\sigma',L^p(\Omega,M_n))$ *for* $\alpha \in \Lambda(T_\sigma',L^p(\Omega,M_n))$.

If $\|\sigma\| = 1$, then the 1-eigenspace $H_1(T_\sigma',L^\infty(\Omega,M_n))$ is a JB*-triple.

Proof. (i) Given that $T_\sigma - \alpha I$ has a bounded inverse $S : L^p(G,M_n) \longrightarrow L^p(G,M_n)$, for some $\alpha \in \mathbb{C}$, we have $Q(T_\sigma - \alpha)Sj = Qj = I_{L^p(\Omega,M_n)}$. Hence $(T_\sigma' - \alpha)QSj = (T_\sigma'Q - \alpha Q)Sj = Q(T_\sigma - \alpha)Sj = I_{L^p(\Omega,M_n)}$ and $\alpha \notin \mathrm{Spec}(T_\sigma',L^p(\Omega,M_n))$.

(ii) This follows from the fact that $T_\sigma'f = \alpha f$ for some $f \in L^p(\Omega,M_n)$ implies that $T_\sigma jf = jT_\sigma'f = \alpha jf \in L^p(G,M_n)$.

(iii) Given $g \in H_\alpha(T_\sigma,L^p(G,M_n))$, we have

$$T_\sigma'Qg = QT_\sigma g = \alpha Qg \in H_\alpha(T_\sigma',L^p(\Omega,M_n)).$$

On the other hand, if $f \in H_\alpha(T_\sigma',L^p(\Omega,M_n))$, then we have $T_\sigma jf = jT_\sigma'f = \alpha jf \in H_\alpha(T_\sigma,L^p(G,M_n))$ and $f = Q(jf)$.

Finally, by Lemma 3.4.2, the 1-eigenspace $H_1(T_\sigma,L^\infty(G,M_n))$ is a JB*-triple for $\|\sigma\| = 1$. If we identify $H_1(T_\sigma',L^\infty(\Omega,M_n))$ as a subspace of $L^\infty(G,M_n)$ via the embedding j, then (iii) implies that it is the range of the contractive projection Q on the JB*-triple $H_1(T_\sigma,L^\infty(G,M_n))$ and hence is itself a JB*-triple by [42]. □

Let $\sigma \in M(G,M_n)$ and $\Omega = G/K$ as above. Motivated by the condition in (3.14), we introduce the following space, for $p = 1, \infty$:

$$H^p(\Omega,M_n) = \{f \in L^p(\Omega,M_n) : \widetilde{jf} * \sigma = \widetilde{jf}, \, jf(\cdot k) = jf(\cdot), \forall k \in K\}. \tag{3.16}$$

Lemma 3.4.5. *Let $\Omega = G/K$ where G is unimodular. Let $\sigma \in M(G,M_n)$ be K-invariant and let $p = 1$ or ∞. Given $f \in H_1(T_\sigma',L^p(\Omega,M_n))$, the function*

$$\widetilde{f}(Kx) := f(Kx^{-1}) \qquad (x \in G)$$

is well-defined and we have $\widetilde{f} \in H^p(\Omega, M_n)$. Further, all maps in the following commutative diagram are surjective:

$$H_1(T_\sigma, L^p(G, M_n)) \xrightarrow{\widetilde{Q}} H^p(\Omega, M_n)$$

$$j \uparrow \qquad\qquad \nearrow \sim$$

$$H_1(T'_\sigma, L^p(\Omega, M_n))$$

where the map $\sim$ is an isometry and $\widetilde{Q}(f) := \widetilde{Q(f)}$. For $p = \infty$, the condition of unimodularity of G can be dropped.

Proof. Let $f \in H_1(T'_\sigma, L^p(\Omega, M_n))$. Given $Kx = Kz$ with $x = kz$ for some $k \in K$, we have

$$f(Kx^{-1}) = f(Kz^{-1}k^{-1})$$
$$= \int_G f(Kz^{-1}k^{-1}y^{-1})d\sigma(y)$$
$$= \int_G f(Kz^{-1}y^{-1})d\sigma(y)$$
$$= f(Kz^{-1})$$

by K-invariance of σ. Hence $\widetilde{f}$ is well defined and $\widetilde{f} \in L^p(G, M_n)$ by unimodularity of G. Moreover, we have $j\widetilde{f} = jf = jf * \sigma$ and $j\widetilde{f}(\cdot k) = j\widetilde{f}(\cdot)$ for all $k \in K$, that is, $\widetilde{f} \in H^p(\Omega, M_n)$.

The map $\sim$ is clearly isometric. To see that it is surjective, pick any $g \in H^p(\Omega, M_n)$ and define $g' : \Omega \longrightarrow M_n$ by

$$g'(Kx) = g(Kx^{-1}) \qquad (x \in G).$$

Then g' is well-defined since $jg(\cdot k) = jg(\cdot)$ for $k \in K$. Also, $g' \in H_1(T'_\sigma, L^p(G, M_n))$ with $\widetilde{g'} = g$.   □

We now give an application to Δ-harmonic functions on Riemannian symmetric spaces. Let $\Omega = G/K$ be a symmetric space of non-compact type. We have already noted that there is no non-zero L^p Δ-harmonic function on Ω, for $1 < p < \infty$, by a result of Yau [64]. However, an analogous result for L^1 Δ-harmonic functions on complete manifolds requires non-negativity of the Ricci curvature (cf. [46]) whereas Ω has non-positive sectional curvature (cf. [37, p.241]). Nevertheless, our previous results can be applied to Ω in this case. The space $H^\infty(\Omega, \mathbb{C})$ below has been defined in (3.16).

Proposition 3.4.6. *Let $\Omega = G/K$ be a symmetric space of non-compact type. Then there is no non-zero L^1 Δ-harmonic function on Ω. The space $H^\infty(\Omega, \mathbb{C})$ contains* *non-constant functions and is exactly the space of bounded Δ-harmonic functions on Ω. Moreover, $H^\infty(\Omega, \mathbb{C})$ is linearly isometric to an abelian C*-algebra.*

Proof. Let $\sigma \in M(G)$ be the absolutely continuous K-invariant *probability* measure introduced in (3.13). By Proposition 3.3.59, $H_1(T_\sigma, L^1(G)) = \{0\}$ since G is not compact. Let $f \in L^1(\Omega)$ be Δ-harmonic. We can choose a bounded approximate identity (u_β) in $C_c^\infty(G)$ such that the net $(jf * u_\beta) L^1$-converges to $jf \in L^1(G)$ with $jf * u_\beta$ bounded on G. The convolution

$$f * u_\beta(Kx) := jf * u_\beta(x) \qquad (x \in G)$$

is a well-defined function on Ω. As in [29, p.367], there is a Laplace operator $\widetilde{\Delta}$ on G satisfying

$$\widetilde{\Delta}(jf) = (\Delta f) \circ q.$$

Hence $jf * u_\beta$ is a bounded $\widetilde{\Delta}$-harmonic function on G, and we have

$$\Delta(f * u_\beta) \circ q = \widetilde{\Delta} j(f * u_\beta) = \widetilde{\Delta}(jf * u_\beta) = 0.$$

Therefore $f * u_\beta$ is a bounded Δ-harmonic function on Ω, and by (3.14), we have

$$((f * u_\beta) \circ q)\widetilde{} * \sigma = ((f * u_\beta) \circ q)\widetilde{}$$

for each β. We note that the semisimple Lie group G is unimodular and hence $g \in L^1(G) \mapsto \widetilde{g} \in L^1(G)$ is an isometry. It follows that $\widetilde{f \circ q} * \sigma = \widetilde{f \circ q}$, that is, $\widetilde{f \circ q} \in H_1(T_\sigma, L^1(G))$ and must be 0. Hence $f = 0$.

Next, the functions in $H^\infty(\Omega, \mathbb{C})$ are all Δ-harmonic on Ω, by the condition (3.14). Conversely, given a bounded Δ-harmonic function f on Ω, then f satisfies (3.14) and we have, by [29, Theorem 4.1],

$$jf(x) = \int_K jf(xky) d\lambda(k)$$

for all $x, y \in G$. It follows that, for each $h \in K$,

$$jf(xh) = \int_K jf(xhke) d\lambda(k) = \int_K jf(xk) d\lambda(k) = jf(x). \qquad (3.17)$$

Therefore $f \in H^\infty(\Omega, \mathbb{C})$.

We now show that $H^\infty(\Omega, \mathbb{C})$ contains non-constant functions. Suppose otherwise. Let $f \in H_1(T_\sigma, L^\infty(G))$. By K-invariance of σ, it is readily verified that the function $F : \Omega \longrightarrow \mathbb{C}$ given by

$$F(Hx) := f(x^{-1}) \qquad (x \in G) \qquad (3.18)$$

is well-defined in $L^\infty(\Omega)$ and satisfies $\widetilde{jF} * \sigma = \widetilde{jF}$. Hence jF is a bounded Δ-harmonic function on Ω and must be constant by assumption. It follows from (3.18) that f is constant. Hence all bounded σ-harmonic functions on G are constant and therefore G must be amenable (cf. [13, Proposition 2.1.3]), contradicting that G/K is of non-compact type.

Finally, since σ is a probability measure, the 1-eigenspace $H_1(T_\sigma, L^\infty(G))$ is an abelian C*-algebra by Lemma 3.4.2. Hence $H_1(T'_\sigma, L^\infty(\Omega))$ is also an abelian C*-algebra since, by Lemma 3.4.4, it is the range of a contractive projection on an abelian C*-algebra. The last assertion then follows from Lemma 3.4.5.   □

Remark 3.4.7. In the setting of Proposition 3.4.6, it has also been shown in [29, p.373] that $H_1(T_\sigma, L^\infty(G))$ is an abelian C*-algebra, by a different method using probability, but it has not been observed in [29] that $H^\infty(\Omega, \mathbb{C})$ is also an abelian C*-algebra.

Remark 3.4.8. We take the opportunity here of noting a missing remark in [14], before [14, Corollary 4.11], namely, that the map $\widetilde{f}(Ha) := f(Ha^{-1})$ there is well-defined for $\Delta f = 0$ because $jf(\cdot k) = jf(\cdot)$ for all k, as in (3.17) above.

Example 3.4.9. The space $H^\infty(\Omega, \mathbb{C})$ is infinite dimensional for the upper half-plane $\Omega = \{z \in \mathbb{C} : \operatorname{Im} z > 0\} = SL(2, \mathbb{R})/SO(2)$ which is of non-compact type.

An application of Proposition 3.4.6 gives an alternative approach to the Poisson representation of harmonic functions on non-compact symmetric spaces given in [29, Theorem 4.2].

Corollary 3.4.10. *Let Δ be a Laplace operator on a symmetric space $\Omega = G/K$ of non-compact type. Then there is a compact Hausdorff space Π with a probability measure μ on Π, and an action $(x, \omega) \in G \times \Pi \mapsto x \cdot \omega \in \Pi$ such that for each bounded Δ-harmonic function f on Ω, there is a unique continuous function $\widehat{f}$ on Π satisfying*

$$f(Kx) = \int_\Pi \widehat{f}(x \cdot \omega) d\mu(\omega) \qquad (x \in G).$$

Proof. By Proposition 3.4.6, the space $H^\infty(\Omega, \mathbb{C})$ of bounded Δ-harmonic functions is an abelian C*-algebra. Let Π be the pure state space of $H^\infty(\Omega, \mathbb{C})$. Then Π is weak* compact Hausdorff and $H^\infty(\Omega, \mathbb{C})$ is isometrically isomorphic to the algebra $C(\Pi)$ of complex continuous functions on Π, via the Gelfand map $f \in H^\infty(\Omega, \mathbb{C}) \mapsto \widehat{f} \in C(\Pi)$, where

$$\widehat{f}(\omega) = \omega(f) \qquad (f \in H^\infty(\Omega, \mathbb{C}), \omega \in \Pi).$$

Define a probability measure $\mu \in C(\Pi)^*$ by

$$\mu(\widehat{f}) = f(K).$$

For each $x \in G$, the right translation $r_x : H^\infty(\Omega, \mathbb{C}) \longrightarrow H^\infty(\Omega, \mathbb{C})$ defined by

$$(r_x f)(Ka) = f(Kax) \qquad (Ka \in \Omega = G/K)$$

is an isometry and induces a surjective linear isometry $\hat{r}_x : C(\Pi) \longrightarrow C(\Pi)$ given by

$$\hat{r}_x(\hat{f}) = \widehat{r_x f}$$

for $f \in H^\infty(\Omega, \mathbb{C})$. Hence, by the Banach-Stone theorem, $\hat{r}_x$ is a composition operator on $C(\Pi)$:

$$\hat{r}_x(\hat{f}) = \hat{f} \circ \varphi_x$$

for some homeomorphism $\varphi_x : \Pi \longrightarrow \Pi$. Since $\varphi_{xy} = \varphi_x \circ \varphi_y$ for all $x, y \in G$, the map

$$(x, \omega) \in G \times \Pi \mapsto x \cdot \omega := \varphi_x(\omega) \in \Pi$$

is an action of G on Π. For every $f \in H^\infty(\Omega, \mathbb{C})$, we have

$$\begin{aligned}
f(Kx) = (r_x f)(K) &= \mu(\widehat{r_x f}) \\
&= \mu(\hat{f} \circ \varphi_x) = \int_\Pi \hat{f}(\varphi_x(\omega)) d\mu(\omega) \\
&= \int_\Pi \hat{f}(x \cdot \omega) d\mu(\omega) \qquad (x \in G).
\end{aligned}$$

$\square$

We now study the JB*-triple $H_1(T_\sigma, L^\infty(G))$ for $\|\sigma\| = 1$. Our main concern is the compatibility of the Jordan structures in $H_1(T_\sigma, L^\infty(G))$ with the ones in its ambient space $L^\infty(G)$. The two structures are different in general. We determine below exactly when they coincide.

A linear subspace V of a JB*-triple Z is called a *subtriple* if it is closed with respect to the triple product in Z which is equivalent to saying that $f \in V$ implies $\{f, f, f\} \in V$, by the polarization identity

$$8\{f, g, h\} = \sum_{\beta^4 = \gamma^2 = 1} \beta\gamma\{(f + \beta g + \gamma h), (f + \beta g + \gamma h), (f + \beta g + \gamma h)\}.$$

Our task is to determine when the eigenspace $H_1(T_\sigma, L^\infty(G))$ is a subtriple of $L^\infty(G)$.

We shall denote $\{f, f, f\}$ by $f^{(3)}$ in any JB*-triple. If a JB*-triple V has a predual which is necessarily unique, then its Jordan triple product is separately weak*-continuous and V contains nonzero *tripotents*, these are the elements $v \in V$ satisfying $v^{(3)} = v$ in which case $\|v\| = 1$.

Lemma 3.4.11. *Let Ω be a locally compact space and let μ be a probability measure on Ω. Let $f \in L^\infty(\Omega, \mu)$ satisfy*

$$\int_\Omega f^{(3)} d\mu = \left(\int_\Omega f d\mu \right)^{(3)} \qquad \text{and} \qquad \left| \int_\Omega f d\mu \right| = \int_\Omega |f| d\mu.$$

Then f is constant μ-almost everywhere.

Proof. If $\int_\Omega |f| d\mu = 0$, there is nothing to prove. We may therefore assume $\int_\Omega |f| d\mu = 1$ by normalizing. The second condition of the lemma implies that

$f = \beta|f|$ μ-almost everywhere, for some complex number β of unit modulus. By the first condition,

$$\int_\Omega f^{(3)} d\mu = \beta \left(\int_\Omega |f| d\mu \right)^{(3)} = \int_\Omega \beta |f|^3 d\mu.$$

We have $2|f|^2 \leq |f|^3 + |f|$ and

$$2 \leq 2 \int_\Omega |f|^2 d\mu \leq \int_\Omega |f|^3 d\mu + \int_\Omega |f| d\mu = 2$$

which gives $2|f|^2 = |f|^3 + |f|$ and hence $|f| = 1$ μ-almost everywhere. $\square$

Example 3.4.12. If $f : \Omega \longrightarrow \mathbb{R}$ satisfies $\int_\Omega f^2 d\mu = (\int_\Omega f d\mu)^2$, then f is constant μ-almost everywhere, but the same conclusion fails if one replaces the square in the integrals by the cube, for instance, $\int_0^1 f^3(x) dx = 0 = \left(\int_0^1 f(x) dx \right)^3$ for $f = \chi_{[0,\frac{1}{2}]} - \chi_{(\frac{1}{2},1]}$.

In what follows, $L^\infty(G)$ is equipped with the Jordan triple product $\{f,h,k\} = f\bar{h}k$ as before.

Lemma 3.4.13. *Let G be a locally compact group and let σ be a complex measure on G with $\|\sigma\| = 1$ and the polar representation $\sigma = \omega \cdot |\sigma|$. The following conditions are equivalent.*

(i) *$H_1(T_\sigma, L^\infty(G))$ is a subtriple of $L^\infty(G)$.*
(ii) *For each $f \in H_1(T_\sigma, L^\infty(G))$, we have $f^{(3)} * \sigma = 0$ λ-a.e. on $f^{-1}(0)$ and $f = \omega(y) f_{y^{-1}}$ λ-a.e. on $G \backslash f^{-1}(0)$, for $|\sigma|$-a.e. y.*

Proof. (i) $\Longrightarrow$ (ii). First, pick an extreme point u in the closed unit ball of $H_1(T_\sigma, L^\infty(G))$. This is possible because $H_1(T_\sigma, L^\infty(G))$ has a predual. By [13, Proposition 2.2.5; p.16], $H_1(T_\sigma, L^\infty(G))$ is the range of a contractive projection P on $L^\infty(G)$ such that $H_1(T_\sigma, L^\infty(G))$ is a JB*-triple in the Jordan triple product

$$[f,h,g] = P\{f,h,k\} \qquad (f,h,k \in H_1(T_\sigma, L^\infty(G)))$$

and also $[f,u,u] = f$. Since $H_1(T_\sigma, L^\infty(G))$ is a subtriple of $L^\infty(G)$, we have

$$f = [f,u,u] = \{f,u,u\} = f|u|^2. \tag{3.19}$$

Since $\|u\|_\infty = 1$, we may assume $|u| \leq 1$ on G, by re-defining u to be 0 on a λ-null set if necessary. Let

$$E = \{x \in G : u^{(3)}(x) = u(x)\overline{u(x)}u(x) = u(x) = u * \sigma(x)\}.$$

Then $\lambda(G \backslash E) = 0$. Choose any $z \in E \cap G \backslash f^{-1}(0)$. We have $|u(z)|^2 = 1$ since $u(z) = u(z)\overline{u(z)}u(z)$. Therefore

$$1 = |u(z)| = \left| \int_G u(zy^{-1})\omega(y)d|\sigma|(y) \right| \le \int_G |u(zy^{-1})\omega(y)||d|\sigma|(y) \le 1.$$

We also have, as $\omega\overline{\omega} = 1$,

$$\int_G (u(zy^{-1})\omega(y))^{(3)}d|\sigma|(y) = u(z) = u^{(3)}(z) = \left(\int_G u(zy^{-1})\omega(y)d|\sigma|(y) \right)^{(3)}.$$

By Lemma 3.4.11, $u(zy^{-1})\omega(y)$ is constant for $|\sigma|$-almost every $y \in G$. Hence $u(z) = u * \sigma(z) = u(zy^{-1})\omega(y)$ for $|\sigma|$-almost every $y \in G$.

Now pick any $f \in H_1(T_\sigma, L^\infty(G))$. Then $f^{(3)} \in H_1(T_\sigma, L^\infty(G))$ by condition (i), and for $f^{(3)}(x) = f(x)\overline{f(x)}f(x) = f^{(3)} * \sigma(x)$, we have $f^{(3)} * \sigma(x) = 0$ if $f(x) = 0$.

Next, condition (i) implies that the function $\{f, f, u\} = |f|^2 u$ belongs to $H_1(T_\sigma, L^\infty(G))$. Let

$$N = E \cap \{x \in G : f(x) = f * \sigma(x) \text{ and } |f|^2(x)u(x) = |f|^2 u * \sigma(x)\}.$$

Then $G \backslash N$ is a λ-null set. Let $x \in N$ and $f(x) \neq 0$. It follows from (3.19) that $u(x) \neq 0$ and hence we have

$$|f|^2(x)u(x) = (|f|^2 u) * \sigma(x) = \int_G |f|^2(xy^{-1})u(xy^{-1})d\sigma(y)$$

$$= u(x) \int_G |f|^2(xy^{-1})d|\sigma|(y)$$

which gives

$$\int_G |f|^2(xy^{-1})d|\sigma|(y) = |f|^2(x) = |f(x)|^2$$

$$= \left| \int_G f(xy^{-1})\omega(y)d|\sigma|(y) \right|^2$$

$$\le \left(\int_G |f(xy^{-1})|d|\sigma|(y) \right)^2$$

$$\le \int_G |f(xy^{-1})|^2 d|\sigma|(y).$$

Hence the above inequalities are equalities and the last one implies that $|f(xy^{-1})|$ is constant for $|\sigma|$-almost every $y \in G$. Also

$$\left| \int_G f(xy^{-1})\omega(y)d|\sigma|(y) \right| = \int_G |f(xy^{-1})|d|\sigma|(y) = \int_G |f(xy^{-1})\omega(y)|d|\sigma|(y)$$

implies that $f(xy^{-1})\omega(y)$ is a constant multiple of $|f(xy^{-1})\omega(y)| = |f(xy^{-1})|$, and hence constant, for $|\sigma|$-almost every $y \in G$, yielding $f(x) = f(xy^{-1})\omega(y)$ for $|\sigma|$-almost every $y \in G$.

(ii) $\implies$ (i). Let $f \in H_1(T_\sigma, L^\infty(G))$. We show $f^{(3)} \in H_1(T_\sigma, L^\infty(G))$. Indeed, for λ-a.e. x in $G \backslash f^{-1}(0)$, we have

$$f^{(3)} * \sigma(x) = \int_G f^{(3)}_{y^{-1}}(x) \omega(y) d|\sigma|(y) = \int_G f^{(3)}_{y^{-1}}(x) \omega^{(3)}(y) d|\sigma|(y)$$
$$= \int_G f^{(3)}(x) d|\sigma|(y) = f^{(3)}(x).$$

For λ-a.e. x in $f^{-1}(0)$, we have

$$f^{(3)}(x) = 0 = f^{(3)} * \sigma(x).$$

$\square$

We note that the eigenspace $H_1(T_\sigma, L^\infty(G))$ is weak* closed in $L^\infty(G)$ and $H_1(T_\sigma, L^\infty(G)) \cap C_b(G)$ is weak* dense in $H_1(T_\sigma, L^\infty(G))$. The latter is a consequence of the existence of a bounded approximate identity (ψ_β) in $L^1(G)$. Given $f \in H_\alpha(T_\sigma, L^\infty(G))$, the convolution $\widetilde{\psi}_\beta * f$ belongs to $H_\alpha(T_\sigma, L^\infty(G)) \cap C_b(G)$, where $\widetilde{\psi}_\beta(x) = \psi_\beta(x^{-1})$, and the net $(\widetilde{\psi}_\beta * f)$ weak* converges to f since, for each $h \in L^1(G)$, we have

$$\langle h, f \rangle = \lim_\beta \langle \psi_\beta * h, f \rangle = \lim_\beta \langle h, \widetilde{\psi}_\beta * f \rangle.$$

Theorem 3.4.14. *Let G be a locally compact group and let σ be an absolutely continuous measure on G with $\|\sigma\| = 1$ and the polar representation $\sigma = \omega \cdot |\sigma|$. The following conditions are equivalent:*

(i) $H_1(T_\sigma, L^\infty(G))$ *is a subtriple of* $L^\infty(G)$;
(ii) $H_1(T_\sigma, L^\infty(G)) = \{f \in L^\infty(G) : f = \omega(y) f_{y^{-1}} \text{ for } |\sigma|\text{-a.e. } y \in G\}$.

Proof. (i) $\implies$ (ii). Given $f \in L^\infty(G)$ satisfying $f = \omega(y) f_{y^{-1}}$ for $|\sigma|$-almost every $y \in G$, it is easily verified that f is σ-harmonic.

Conversely, for each $f \in H_1(T_\sigma, L^\infty(G))$, absolute continuity of σ implies that $f * \sigma \in C_b(G)$ and that we may assume, by re-defining if necessary, that $f = f * \sigma$ on G. By Lemma 3.4.13, we have $f(x) = \omega(y) f(xy^{-1})$ for $|\sigma|$-a.e. $y \in G$ if $f(x) \neq 0$.

Suppose $f(x) = 0$. Then $_a f(x) \neq 0$ for some left translate $_a f$ of f. We have $_a f \in H_1(T_\sigma, L^\infty(G))$, and $|f|^2 {}_a f = \{f, f, {}_a f\} \in H_1(T_\sigma, L^\infty(G))$ by condition (i). Hence

$$0 = |f|^2 {}_a f(x) = |f|^2 {}_a f * \sigma(x)$$
$$= \int_G |f|^2 (xy^{-1}) f(axy^{-1}) \omega(y) d|\sigma|(y)$$
$$= f(ax) \int_G |f(xy^{-1})|^2 d|\sigma|(y)$$

which implies $|f(xy^{-1})| = 0$ as well as $f(x) = 0 = f(xy^{-1}) \omega(y)$ for $|\sigma|$-almost every $y \in G$.

(ii) $\implies$ (i). If $f \in H_1(T_\sigma, L^\infty(G))$, then condition (ii) implies that $f^{(3)}$ is also in $H_1(T_\sigma, L^\infty(G))$. Hence $H_1(T_\sigma, L^\infty(G))$ is a subtriple of $L^\infty(G)$.

$\square$

If σ is a probability measure on G, then $H_1(T_\sigma, L^\infty(G))$ contains constant functions and if it is a subtriple of $L^\infty(G)$, then it is also a $*$-subalgebra, and *vice versa*, since $\{f, \mathbf{1}, h\} = fh$ and $f^* = \{\mathbf{1}, f, \mathbf{1}\}$ in $L^\infty(G)$.

Corollary 3.4.15. *Let σ be an absolutely continuous probability measure on a locally compact group G. The following conditions are equivalent:*

(i) $H_1(T_\sigma, L^\infty(G))$ *is a $*$-subalgebra of $L^\infty(G)$;*
(ii) $H_1(T_\sigma, L^\infty(G))$ *is a subtriple of $L^\infty(G)$;*
(iii) $H_1(T_\sigma, L^\infty(G)) = \{f \in L^\infty(G) : f = f_{a^{-1}} \ \forall a \in \operatorname{supp}\sigma\}.$

Proof. (ii) $\implies$ (iii). Let $f \in H_1(T_\sigma, L^\infty(G))$. By absolute continuity of σ, we may take f to be continuous. By Theorem 3.4.14, the open set $\{y \in G : f \neq f_{y^{-1}}\}$ is disjoint from $\operatorname{supp}\sigma$. $\qquad\square$

Chapter 4
Convolution Semigroups

In this Chapter, we study matrix harmonic functions and contractivity properties of a semigroup of matrix convolution operators $\{T_{\sigma_t}\}_{t>0}$ where $\|\sigma_t\| = 1$. We show that, for $1 < p \leq \infty$, there is a contractive projection $P : L^p(G, M_n) \longrightarrow L^p(G, M_n)$ whose range is the intersection of 1-eigenspaces of $\{T_{\sigma_t}\}$:

$$\bigcap_{t>0} H_1(T_{\sigma_t}, L^p(G, M_n)) = \{f \in L^p(G, M_n) : f = f * \sigma_t \text{ for all } t > 0\}.$$

This is the space of matrix L^p harmonic functions for the generator of the semigroup, and it is a JB*-triple if $p = \infty$, in which case it is nontrivial if σ_t are positive and G non-amenable.

If $\mathcal{L}$ is a second order G-invariant elliptic operator on a connected Lie group G, annihilating constant functions, or more generally, a translation invariant Dirichlet form on a locally compact group G, then it generates a convolution semigroup $\{T_{\sigma_t}\}_{t>0}$ and our results can be applied to this setting. For instance, one can derive a Poisson representation for L^∞ $\mathcal{L}$-harmonic functions on G, and show that all L^p $\mathcal{L}$-harmonic functions are constant for $1 \leq p < \infty$.

We study hypercontractivity of the semigroup $\{T_{\sigma_t}\}$ in the last part of this Chapter. We show that Gross's seminal result on hypercontractivity and log-Sobolev inequality can be extended to the matrix setting.

4.1 Harmonic Functions of Semigroups

Let G be a connected Lie group and let $\mathcal{L}$ be a second order G-invariant elliptic differential operator on G, annihilating the constant functions. A C^2-function on G is $\mathcal{L}$-*harmonic* if $\mathcal{L}f = 0$. By [39, Theorem 5.1], $\mathcal{L}$ generates a convolution semigroup of absolutely continuous probability measures $\{\sigma_t\}_{t>0}$ on G, giving rise to a semigroup $T_t : L^p(G) \longrightarrow L^p(G)$ of convolution operators

$$T_t(f) = f * \sigma_t \quad (t > 0)$$

C.-H. Chu, *Matrix Convolution Operators on Groups*. Lecture Notes in Mathematics 1956, 87
doi: 10.1007/978-3-540-69798-5, © Springer-Verlag Berlin Heidelberg 2008

(cf. [1, p.134]). The intersection of eigenspaces

$$\bigcap_{t>0} H_1(T_{\sigma_t}, L^\infty(G)) = \{f \in L^\infty(G) : f * \sigma_t = f \text{ for all } t > 0\}$$

is the space of L^∞ $\mathcal{L}$-harmonic functions on G (cf. [1, Proposition V.6] and [25, Theorem 5.9]). More generally, for any locally compact group G, if a self-adjoint operator $\mathcal{L}$ in $L^2(G)$ is a Dirichlet form (cf. [21]) and if $\mathcal{L}$ commutes with left translations, then it generates a convolution semigroup $e^{-t\mathcal{L}} : L^p(G) \longrightarrow L^p(G)$.

In this section we study convolution semigroups $\{\sigma_t\}_{t>0}$ of matrix-valued measures on a locally compact group G and our focus is on the harmonic functions of the semigroup, namely, the intersection of eigenspaces:

$$\bigcap_{t>0} H_1(T_{\sigma_t}, L^p(G, M_n)) = \{f \in L^p(G, M_n) : f = f * \sigma_t \text{ for all } t > 0\}.$$

We show that it is the range of a contractive projection on $L^p(G, M_n)$ and the space $\bigcap_{t>0} H_1(T_{\sigma_t}, L^\infty(G, M_n))$ carries the structure of a Jordan triple system. We show how these results can be applied to $\mathcal{L}$-harmonic functions on Lie groups.

By a *(one-parameter) convolution semigroup of M_n-valued measures* on a locally compact group G with identity e, we mean a family $\{\sigma_t\}_{t>0}$ of measures in $M(G, M_n)$ satisfying

(i) $\|\sigma_t\| = 1$,
(ii) $\sigma_s * \sigma_t = \sigma_{s+t}$,
(iii) $\delta_e = \text{w}^*\text{-}\lim_{t\downarrow 0} \sigma_t$

where the weak* topology on $M(G, M_n)$ is defined by the duality, as in [9, Lemma 6],

$$\langle f, \mu \rangle = \text{Tr} \int_G f d\mu \qquad (f \in C_0(G, M_n^*), \mu \in M(G, M_n)).$$

We note that, if $\{\sigma_t\}_{t>0}$ are *probability* measures, then condition (iii) above is equivalent to the following condition in [39] for a convolution semigroup:

(iv) $\lim_{t\downarrow 0} \sigma_t(V) = 1$ for every open set V containing e

in which case, we have

$$f(e) = \lim_{t\downarrow 0} \langle f, \sigma_t \rangle \qquad \text{for each } f \in C_b(G). \tag{4.1}$$

Remark 4.1.1. Although one could adopt the weaker condition $\|\sigma_t\| \leq 1$ for (i) above, we use $\|\sigma_t\| = 1$ instead for the purpose of discussing harmonic functions.

Let $\{\sigma_t\}_{t>0}$ be a convolution semigroup of M_n-valued measures on G. Given $h \in C_c(G, M_n)$, we have

$$\langle g, h * \sigma_t \rangle = \langle \widetilde{g} * h, \widetilde{\sigma}_t \rangle \qquad (g \in C_c(G, M_n))$$

which implies that $(h * \sigma_t)$ weakly converges to h in $L^p(G, M_n)$ as $t \downarrow 0$, for $1 < p < \infty$. It follows that $\{\sigma_t\}_{t>0}$ generates a *strongly continuous* contractive semigroup of convolution operators

$$T_0 = I, \quad T_t : f \in L^p(G, M_n) \mapsto f * \sigma_t \in L^p(G, M_n)$$

for $1 < p < \infty$, where $f = \text{w-}\lim_{t \downarrow 0} T_t f$ since, for each $g \in C_c(G, M_n)$, we have

$$|\langle T_t f, g \rangle - \langle f, g \rangle| \leq |\langle T_t f, g \rangle - \langle T_t h, g \rangle| + |\langle T_t h, g \rangle - \langle h, g \rangle| + |\langle h, g \rangle - \langle f, g \rangle|$$
$$\leq 2\|f - h\| \|g\| + |\langle h * \sigma_t, g \rangle - \langle h, g \rangle|$$

which can be made arbitrarily small for $t \downarrow 0$ by choosing $h \in C_c(G, M_n)$.

If $\{\sigma_t\}_{t>0}$ are probability measures, the semigroup $\{T_t\}_{t \geq 0} : L^1(G) \longrightarrow L^1(G)$ is also strongly continuous by (4.1).

Example 4.1.2. For a family $\{\sigma_t\}_{t>0}$ of measures, the condition that $\lim_{t \downarrow 0} |\sigma_t|(V) = 1$, for every open set V containing e, is not sufficient to yield $f = \text{w-}\lim_{t \downarrow 0} T_t f$ in $L^p(G, M_n)$. Let $\{\sigma_t\}_{t>0}$ be a family of signed measures on $\mathbb{R}$ defined by

$$d\sigma_t(x) = \frac{1}{t^2} \varphi_t(x) dx$$

where $t > 0$ and

$$\varphi_t(x) = \begin{cases} x \text{ for } -t < x < t \\ 0 \text{ otherwise.} \end{cases}$$

Then $\|\sigma_t\| = 1$ and $\lim_{t \downarrow 0} |\sigma_t|(V) = 1$ for every open interval V containing 0. Let $f = \chi_{(0,1)} \in L^1(\mathbb{R})$ be the characteristic function of $(0,1)$. For $t < 1/2$, we have

$$T_t f(x) = (f * \sigma_t)(x) = \begin{cases} \dfrac{x^2 - t^2}{2t^2} & \text{for } -t \leq x \leq t, \\[2mm] \dfrac{2x - x^2 + t^2 - 1}{2t^2} & \text{for } 1 - t \leq x \leq 1 + t, \\[2mm] 0 & \text{otherwise} \end{cases}$$

and $\langle T_t f, \chi_{(0, \frac{1}{2})} \rangle = -t/3 \nrightarrow \langle f, \chi_{(0, \frac{1}{2})} \rangle$ in $L^1(\mathbb{R})$ as $t \downarrow 0$.

Lemma 4.1.3. *Let $1 \leq p < \infty$ and let $T_t : L^p(G, M_n) \longrightarrow L^p(G, M_n)$ be a strongly continuous one-parameter semigroup of bounded operators, with generator $\mathcal{L}_p$ and domain $\text{Dom}(\mathcal{L}_p) \subset L^p(G, M_n)$. Let $f \in L^p(G, M_n)$ and $\alpha \in \mathbb{C}$. The following conditions are equivalent:*

(i) $T_t f = e^{\alpha t} f$ $(t > 0)$;
(ii) $f \in \text{Dom}(\mathcal{L}_p)$ and $\mathcal{L}_p f = \alpha f$.

Proof. (i) $\Longrightarrow$ (ii). We have

$$\lim_{t\downarrow 0}\frac{1}{t}(T_t f - f) = \lim_{t\downarrow 0}\frac{e^{\alpha t}-1}{t}f = \alpha f.$$

Therefore $f \in \mathrm{Dom}(\mathcal{L}_p)$ and $\mathcal{L}_p f = \alpha f$.

(ii) $\Longrightarrow$ (i). We have $T_t f - f = \int_0^t T_x \mathcal{L}_p f\, dx = \alpha \int_0^t T_x f\, dx$. Hence

$$\frac{d}{dt}T_t f = \alpha T_t f$$

which gives $T_t f = e^{\alpha t}f$. $\qquad\square$

Let $\mathcal{S} = \{\sigma_t : t > 0\}$ be a convolution semigroup of M_n-valued measures on G. We consider the semigroup of convolution operators $T_t : L^p(G,M_n) \longrightarrow L^p(G,M_n)$, where $1 \leq p \leq \infty$, defined by

$$T_0 = I, \qquad T_t(f) = f * \sigma_t \qquad (t > 0).$$

A function $f \in L^p(G,M_n)$ is called $\mathcal{S}$-*harmonic* or $(\sigma_t)_{t>0}$-*harmonic* if $f = f * \sigma_t$ in $L^p(G,M_n)$ for all $t > 0$. Let

$$H_{\mathcal{S}}^p(G,M_n) = \{f \in L^p(G,M_n) : f = f * \sigma_t \text{ for all } t > 0\} = \bigcap_{t>0} H_1(T_{\sigma_t}, L^p(G,M_n))$$

be the Banach space of M_n-valued $\mathcal{S}$-harmonic L^p functions on G. Let $\mathcal{L}_p$ be the generator of $\{T_t\}_{t\geq 0} : L^p(G,M_n) \longrightarrow L^p(G,M_n)$ for $1 < p < \infty$. Then

$$H_{\mathcal{S}}^p(G,M_n) = \{f \in \mathrm{Dom}(\mathcal{L}_p) : \mathcal{L}_p f = 0\}$$

by Lemma 4.1.3. Since $\{T_t\}_{t\geq 0}$ is contractive, Lemma 4.1.3 implies that, if α is an eigenvalue of $\mathcal{L}_p$, then $|\exp(\alpha t)| \leq 1$ for all $t > 0$ and in particular, $\mathrm{Re}\,\alpha \leq 0$.

We define $\widetilde{\mathcal{S}} = \{\widetilde{\sigma}_t : t > 0\}$ where $d\widetilde{\sigma}_t(x) = d\sigma_t(x^{-1})$. By (3.3), $\widetilde{\mathcal{S}}$ is also a one-parameter convolution semigroup of measures on G, with respect to the convolution $*_\ell$.

The following result extends Proposition 3.3.56, with analogous proof. We outline the main steps of the arguments.

Proposition 4.1.4. *Let $1 < p \leq \infty$ and let $\mathcal{S} = \{\sigma_t\}_{t>0}$ be a convolution semigroup of M_n-valued measures on G. Then there is a contractive projection $P_{\mathcal{S}} : L^p(G,M_n) \longrightarrow L^p(G,M_n)$ with range $H_{\mathcal{S}}^p(G,M_n)$ and $P_{\mathcal{S}}$ commutes with left translations. Further, for $1 < p < \infty$, the projection $P_{\mathcal{S}}$ is the dual map of a contractive projection $Q_{\widetilde{\mathcal{S}}} : L^q(G,M_n) \longrightarrow L^q(G,M_n)$ and $H_{\mathcal{S}}^p(G,M_n) = \widetilde{H}_{\widetilde{\mathcal{S}}}^q(G,M_n)^*$ where*

$$\widetilde{H}_{\widetilde{\mathcal{S}}}^q(G,M_n) = \bigcap_{t>0} H_1(L_{\widetilde{\sigma}_t}, L^p(G,M_n)).$$

Proof. For each $t > 0$, let $T_t : L^p(G, M_n) \longrightarrow L^p(G, M_n)$ be the convolution operator

$$T_t(f) = f * \sigma_t.$$

We have $\|T_t\| \leq 1$ and T_t is weakly continuous when $L^p(G, M_n)$ is equipped with the weak topology. Let $\mathcal{K} = \overline{co}\{T_t : t > 0\}$ be the closed convex hull of $\{T_t : t > 0\}$ with respect to the product topology $\mathcal{T}$ of $L^p(G, M_n)^{L^p(G, M_n)}$ where $L^p(G, M_n)$ is equipped with the weak topology. Then $\mathcal{K}$ is compact. Define $\Phi_t : \mathcal{K} \longrightarrow \mathcal{K}$ by

$$\Phi_t(\Lambda)(f) = \Lambda(f) * \sigma_t \qquad (\Lambda \in \mathcal{K}, f \in L^p(G, M_n)).$$

It is straightforward to verify that Φ_t is well-defined and $\mathcal{T}$-continuous. Since $\sigma_s * \sigma_t = \sigma_t * \sigma_s$, the family $\{\Phi_t\}_{t>0}$ is a commuting family of continuous affine maps on $\mathcal{K}$ and by the Markov-Kakutani fixed-point theorem (cf. [24, p. 456]), $\{\Phi_t\}_{t>0}$ has a common fixed-point $P_S \in \mathcal{K}$ which is the required contractive projection.

The projection $Q_{\widetilde{S}}$ is constructed similarly via the maps

$$g \in L^q(G, M_n) \mapsto \Psi_t(\widetilde{\Lambda})(g) = \widetilde{\sigma}_t *_\ell \widetilde{\Lambda}(g) \in L^q(G, M_n)$$

with $\widetilde{\Lambda} \in \widetilde{\mathcal{K}} = \overline{co}\{\widetilde{T}_t : t > 0\}$ and $\widetilde{T}_t(g) = \widetilde{\sigma}_t *_\ell g$.

The proof of $P_S = Q_{\widetilde{S}}^*$ is similar to the arguments for Proposition 3.3.56, noting that $P_S(f * \sigma_t) = P_S(f) * \sigma_t = P_S(f)$ for each $f \in L^p(G, M_n)$ and $Q_{\widetilde{S}}(\widetilde{\sigma}_t *_\ell g) = \widetilde{\sigma}_t *_\ell Q_{\widetilde{S}}(g) = Q_{\widetilde{S}}(g)$ for each $g \in L^q(G, M_n)$.

Finally, we have $\widetilde{H}_{\widetilde{S}}^q(G, M_n)^* = Q_{\widetilde{S}}(L^q)^* \simeq L^p/P_{\widetilde{S}}(L^q)^\perp \simeq H_S^p(G, M_n)$ where $L^q = L^q(G, M_n)$. $\qquad\qquad\square$

Remark 4.1.5. By the above construction of P_S, there is a net of measures (μ_α) in the convex hull of $\{\sigma_t : t > 0\}$ such that

$$P_S(f) = \text{w}^*\text{-}\lim_\alpha f * \mu_\alpha$$

for every $f \in L^\infty(G, M_n)$.

Corollary 4.1.6. *Then space $H_S^\infty(G, M_n)$ is a JBW*-triple.*

Proof. Since $H_S^\infty(G, M_n)$ is the range of the contractive projection $P_S : L^\infty(G, M_n) \longrightarrow L^\infty(G, M_n)$ in Proposition 4.1.4, it is a JB*-triple with the following Jordan triple product

$$\{f, g, h\} = \frac{1}{2} P_S(fg^*h + hg^*f).$$

As the map $f \in L^\infty(G, M_n) \mapsto f * \sigma_t \in L^\infty(G, M_n)$ is weak* continuous, $H_S^\infty(G, M_n)$ is weak* closed in $L^\infty(G, M_n)$ and has a predual, that is, it is a JBW*-triple. $\qquad\square$

As before, let $\mathbf{1} : G \longrightarrow M_n$ be the constant function with value $I \in M_n$.

Proposition 4.1.7. *Let* $S = \{\sigma_t\}_{t>0}$ *be a convolution semigroup of positive* M_n-*valued measures on a locally compact group* G *such that* $\{0\} \neq H_S^\infty(G,M_n) \subset M_n\mathbf{1}$. *Then* G *is amenable.*

Proof. If $f \in H_S^\infty(G,M_n)$ and $f = A\mathbf{1}$ for some $A \in M_n$, then we have $A\mathbf{1} = f * \sigma_t = A\sigma_t(A)\mathbf{1}$ for all $t > 0$. Hence

$$H_S^\infty(G,M_n) = \{A\mathbf{1} : A \in M_n, A = A\sigma_t(G), \forall t > 0\}.$$

Let $P_S : L^\infty(G,M_n) \longrightarrow H_S^\infty(G,M_n)$ be the contractive projection in Proposition 4.1.4 and let $P_S(\mathbf{1}) = A\mathbf{1}$. Then $\|A\| \leq 1$ and by Remark 4.1.5, we have $A \geq 0$ in M_n and $A \neq 0$, for if $0 \neq f = B\mathbf{1} \in H_S^\infty(G,M_n)$, then $BP_S(\mathbf{1}) = B\mathbf{1}$.

We can find a state φ of $L^\infty(G,M_n)$ such that $\varphi(P_S(\mathbf{1})) = 1$ (cf. [62, p.130]). Since P_S commutes with left translations, we have $P_S\ell_x(f) = \ell_xP_S(f) = P_S(f)$ for all $f \in L^\infty(G)$ and $x \in G$, where $P_S(f)$ is a constant function. It follows that $\varphi \circ P$ is a left-invariant state of $L^\infty(G,M_n)$ and the function

$$m : h \in L^\infty(G) \mapsto \varphi(P(h \otimes I)) \in \mathbb{C}$$

is a left-invariant mean. Hence G is amenable. $\square$

We now apply the above results to the heat semigroup on a Lie group. We write $H_S^p(G)$ for $H_S^p(G,\mathbb{C})$. Let G be a connected Lie group and let $\mathcal{L}$ be a second order G-invariant elliptic differential operator on G, annihilating the constant functions. Then $\mathcal{L}$ generates a convolution semigroup $\{\sigma_t\}_{t>0}$ of absolutely continuous probability measures on G, giving rise to a strongly continuous one-parameter semigroup $(T_t)_{t\geq 0}$ of convolution operators on $L^p(G)$ for $1 \leq p < \infty$. The generator $\mathcal{L}_p$ of $(T_t)_{t\geq 0}$ in $L^p(G)$ coincides with $\mathcal{L}$ on $C_c^\infty(G)$. Our first application is the following uniqueness result.

Proposition 4.1.8. *Let* G *be a connected Lie group. For* $1 \leq p < \infty$, *all* L^p $\mathcal{L}$-*harmonic functions on* G *are constant.*

Proof. Let $S = \{\sigma_t\}_{t>0}$ be the semigroup of absolutely continuous probability measures generated by $\mathcal{L}$. Then $H_S^p(G)$ contains the space of L^p $\mathcal{L}$-harmonic functions on G. Since G is connected, each σ_t is adapted and by [10, Theorem 3.1], we have $H_S^p(G) \subset H_1(T_{\sigma_t}, L^p(G)) \subset \mathbb{C}\mathbf{1}$. $\square$

It has been shown by Yau [64] (see also [32]) that all L^p Δ-harmonic functions on a complete Riemannian manifold are constant, for $1 < p < \infty$, where Δ is the Laplace operator of the Riemannian metric of the manifold. This result is false for $p = 1, \infty$, but is true if in addition, the manifold has non-negative Ricci curvature [46, 65]. As shown by Milnor [50, Theorem 2.5], almost any Lie group admits a left invariant Riemannian metric for which the Ricci curvature changes sign. Although the L^1 result for manifolds cannot be applied directly to all Lie groups, it follows from Proposition 4.1.8 that all L^1 Δ-harmonic functions on Lie groups are constant. For the case $p = \infty$, Proposition 4.1.7 gives an alternative proof for the amenability of a Lie group if all bounded Δ-harmonic functions on it are constant.

As another application of Proposition 4.1.4, we give a Poisson representation of bounded $\mathcal{L}$-harmonic functions on Lie groups, similar to the construction in Corollary 3.4.10. Let $\mathcal{S} = (\sigma_t)_{t>0}$ be the semigroup generated by $\mathcal{L}$. The functions in $H_{\mathcal{S}}^{\infty}(G)$ are exactly the bounded $\mathcal{L}$-harmonic functions on G. Since $\mathbf{1}$ is an extreme point of the unit ball of $H_{\mathcal{S}}^{\infty}(G)$, by Corollary 4.1.6, $H_{\mathcal{S}}^{\infty}(G)$ is a an abelian von Neumann algebra with product and involution:

$$f \cdot g = P_{\mathcal{S}}(fg), \qquad f^* = P_{\mathcal{S}}(\bar{f}) = \text{w}^*\text{-}\lim_{\alpha} \bar{f} * \mu_\alpha = \overline{P_{\mathcal{S}}(f)} = \bar{f}$$

where $\mu_\alpha \in \text{co}\{\sigma_t : t > 0\}$ is a probability measure.

Let Ω be the pure state space of $H_{\mathcal{S}}^{\infty}(G)$. Then Ω is weak* compact Hausdorff and $H_{\mathcal{S}}^{\infty}(G)$ is isometrically isomorphic to the algebra $C(\Omega)$ of complex continuous functions on Ω, via the Gelfand map $f \in H_{\mathcal{S}}^{\infty}(G) \mapsto \hat{f} \in C(\Omega)$, where

$$\hat{f}(\omega) = \omega(f) \qquad (f \in H_{\mathcal{S}}^{\infty}(G), \omega \in \Omega).$$

Proposition 4.1.9. *Let $\mathcal{L}$ be a second order G-invariant elliptic differential operator on a connected Lie group G, annihilating the constant functions. Then there is a compact Hausdorff space Ω with a probability measure μ on Ω, and an action $(\omega, x) \in \Omega \times G \mapsto \omega \cdot x \in \Omega$ such that for each bounded $\mathcal{L}$-harmonic function f on G, there is a unique complex continuous function $\hat{f}$ on Ω such that*

$$f(x) = \int_\Omega \hat{f}(\omega \cdot x) d\mu(\omega) \qquad (x \in G).$$

Proof. Let $\mathcal{S}$ be the semigroup of probability measures on G generated by $\mathcal{L}$. Let Ω be the pure state space of $H_{\mathcal{S}}^{\infty}(G)$. Define a probability measure $\mu \in C(\Omega)^*$ by

$$\mu(\hat{f}) = f(e)$$

where $f \in H_{\mathcal{S}}^{\infty}(G) \mapsto \hat{f} \in C(\Omega)$ is the above Gelfand map and e is the identity of G.

For each $x \in G$, the left translation $\ell_x : H_{\mathcal{S}}^{\infty}(G) \longrightarrow H_{\mathcal{S}}^{\infty}(G)$ induces a surjective linear isometry $\hat{\ell}_x : C(\Omega) \longrightarrow C(\Omega)$ given by

$$\hat{\ell}_x(\hat{f}) = \widehat{\ell_x f}$$

for $f \in H_{\mathcal{S}}^{\infty}(G)$. Hence $\hat{\ell}_x$ is a composition operator on $C(\Omega)$:

$$\hat{\ell}_x(\hat{f}) = \hat{f} \circ \varphi_x$$

for some homeomorphism $\varphi_x : \Omega \longrightarrow \Omega$. Since $\varphi_{xy} = \varphi_x \circ \varphi_y$ for all $x, y \in G$, the map

$$(\omega, x) \in \Omega \times G \mapsto \omega \cdot x := \varphi_{x^{-1}}(\omega) \in \Omega$$

is a (right) action of G on Ω. For every $f \in H_{\mathcal{S}}^{\infty}(G)$, we have

$$f(x) = \ell_{x^{-1}}f(e) = \mu(\widehat{\ell_{x^{-1}}f})$$

$$= \mu(\widehat{f} \circ \varphi_{x^{-1}}) = \int_\Omega \widehat{f}(\varphi_{x^{-1}}(\omega))d\mu(\omega)$$

$$= \int_\Omega \widehat{f}(\omega \cdot x)d\mu(\omega) \qquad (x \in G).$$

□

Example 4.1.10. An unbounded self-adjoint positive operator $\mathcal{L}$ on $L^2(G)$ is called a Dirichlet form if it satisfies the Beurling-Deny conditions as in [21]. A concrete example is the second order elliptic operator $\mathcal{L}$ discussed above. The operator $-\mathcal{L}$ generates a contractive semigroup $e^{-t\mathcal{L}} : L^2(G) \longrightarrow L^2(G)$ which can be extended to a contractive semigroup $T_p(t) : L^p(G) \longrightarrow L^p(G)$. If $\mathcal{L}$ commutes with left translations, that is, if the left translation $\ell_x f$ lies in the domain of $\mathcal{L}$ for every f in the domain and $x \in G$, and $\mathcal{L}\ell_x f = \ell_x \mathcal{L}f$, then $e^{-t\mathcal{L}}$, and also $T_p(t)$, commutes with left translations and hence $T_1(t) : L^1(G) \longrightarrow L^1(G)$ is a convolution semigroup $T_1(t) = T_{\sigma_t}$ by Corollary 3.1.11. Further, the contractivity of $e^{-t\mathcal{L}}$ on $L^\infty(G)$ implies that $T_p(t) = T_{\sigma_t}$ on L^p for all $p < \infty$, by Proposition 3.1.10. Hence the above results on semigroups, for instance, Proposition 4.1.4, can be applied to $T_p(t)$ and the $\mathcal{L}_p$-harmonic functions. The L^p-spectrum $\mathrm{Spec}\,(\mathcal{L}_p, L^p)$ of the generator $\mathcal{L}_p$ satisfies $\exp(t\,\mathrm{Spec}\,(\mathcal{L}_p, L^p)) \subset \mathrm{Spec}\,(T_{\sigma_t}, L^p)$.

4.2 Hypercontractivity

We now discuss hypercontractivity of convolution semigroups $T_t = T_{\sigma_t} : L^p(G, M_n) \longrightarrow L^p(G, M_n)$. We are concerned with the question of '*smoothing*' of the semigroup $\{T_t\}_{t>0}$, that is, the question of contractivity of T_t as an operator from $L^p(G, M_n)$ to $L^q(G, M_n)$ for some $q > p$. We extend Gross's seminal result in [36] on hypercontractive semigroups to this setting. For this purpose, the appropriate norm to use for M_n is the Hilbert-Schmidt norm and the setting for the remaining section will be that of the spaces $L^p(G, (M_n, \| \cdot \|_{hs}))$ which will be denoted by $L^p(G, M_{n,2})$ to simplify notation. Recall that $L^2(G, M_{n,2})$ is equipped with the inner product

$$\langle f, g \rangle_2 = \int_G \mathrm{Tr}\,(f(x)g(x)^*)d\lambda(x).$$

If there is no confusion, we write $\langle \cdot, \cdot \rangle$ for $\langle \cdot, \cdot \rangle_2$.

Given a measure $\sigma \in M(G, M_n)$, we define its *adjoint* $\sigma^* \in M(G, M_n)$ by

$$\sigma^*(E) = \sigma(E)^* \in M_n$$

for each Borel set $E \subset G$. Recall that $d\widetilde{\sigma}(x) = d\sigma(x^{-1})$.

We first discuss positivity of the semigroup $\{T_t\}$. Let $-\mathcal{L}$ be the generator of $\{T_t\}_{t \geq 0}$ in $L^2(G, M_{n,2})$. For $f, h \in L^2(G, M_{n,2})$, we have

$$\int_G h(f * \tilde{\sigma}^*)^* d\lambda = \int_G (h * \sigma) f^* d\lambda.$$

It follows that the domain of the adjoint of $-\mathcal{L}$ is given by

$$D(-\mathcal{L}^*) = \left\{ f \in L^2(G, M_{n,2}) : \lim_{t \downarrow 0} \frac{1}{t}(f * \tilde{\sigma}_t^* - f) \text{ exists in } L^2(G, M_{n,2}) \right\}$$

and $\mathcal{L}$ is self-adjoint if, and only if, $\tilde{\sigma}_t^* = \sigma_t$ for all $t > 0$.

Since $\|\sigma_t\| = 1$, we have $\|T_t\|_2 \leq 1$ for all $t > 0$ and hence $\mathcal{L}$ is a *positive* operator, that is,

$$\langle \mathcal{L}f, f \rangle \geq 0$$

for each f in the domain $D(\mathcal{L}) \subset L^2(G, M_{n,2})$ of $\mathcal{L}$. Therefore the operator $\mathcal{L}^{1/2}$ is well-defined. We define the *quadratic form of* $\mathcal{L}$ to be the quadratic form Q with domain $D(\mathcal{L})$, given by

$$Q(f, g) = \langle \mathcal{L}f, g \rangle = \langle \mathcal{L}^{1/2} f, \mathcal{L}^{1/2} g \rangle \qquad (f, g \in D(\mathcal{L}))$$

where we use the same symbol Q for the associated symmetric bilinear form.

Let $M_n^+ = \{A^*A : A \in M_n\}$ be the positive cone in the C*-algebra M_n. We call a function $f \in L^p(G, M_{n,2})$ *positive* and denote this by $f \geq 0$, if $f(x) \in M_n^+$ for λ-almost all $x \in G$. Given a function $f : G \longrightarrow M_n$, we define the functions $f^*, |f| :$ $G \longrightarrow M_n$ by

$$f^*(x) = f(x)^* \quad \text{and} \quad |f|(x) = |f(x)| = (f(x)f(x)^*)^{1/2} \quad (x \in G).$$

If $f^* = f$, we define the *positive* and *negative* parts of f by $f^+(x) = f(x)^+$ and $f^-(x) = f(x)^-$ respectively.

A map $T : L^p(G, M_{n,2}) \longrightarrow L^p(G, M_{n,2})$ is called *positivity preserving*, in symbol, $T \geq 0$, if $f \geq 0$ implies $Tf \geq 0$ for $f \in L^p(G, M_{n,2})$. Since $C_c(G, M_{n,2}) \subset L^p(G, M_{n,2})$, a convolution operator $T_\sigma : L^p(G, M_{n,2}) \longrightarrow L^p(G, M_{n,2})$ is positivity preserving if, and only if, $\sigma \geq 0$.

A semigroup $\{T_t\}_{t \geq 0}$ of operators on $L^p(G, M_{n,2})$ is called *positive* if $T_t \geq 0$ for all $t > 0$. The semigroup $\{T_t\}_{t \geq 0}$ induced by a convolution semigroup $\{\sigma_t\}_{t > 0}$ of M_n-valued measures on G is positive if, and only if, $\sigma_t \geq 0$ for all $t > 0$. The following conditions for the positivity of a semigroup $\{T_t\}_{t \geq 0}$ in terms of its generator $\mathcal{L}$ in $L^2(G, M_{n,2})$ are well-known in the scalar case. The proof for the matrix-valued case is similar to [21, Theorem 1]. We note that, for a positive operator $\mathcal{L}$ in a Hilbert space H and for $\alpha > 0$, the operator $\alpha + \mathcal{L}$ has a bounded inverse on H.

Proposition 4.2.1. *Let $\mathcal{L}$ be a self-adjoint positive operator in $L^2(G, M_{n,2})$ and let $-\mathcal{L}$ generate a semigroup $\{T_t\}_{t \geq 0}$ of operators on $L^2(G, M_{n,2})$. Let Q be the quadratic form of $\mathcal{L}$. The following conditions are equivalent.*

(i) *$T_t \geq 0$ for $t > 0$.*
(ii) *Given $\varphi = \varphi^* \in D(\mathcal{L}^{1/2})$, we have $|\varphi| \in D(\mathcal{L}^{1/2})$ and $Q(|\varphi|) \leq Q(\varphi)$.*
(iii) *Given $\varphi = \varphi^* \in D(\mathcal{L}^{1/2})$, we have $|\varphi| \in D(\mathcal{L}^{1/2})$ and $Q(\varphi^+, \varphi^-) \leq 0$.*

(iv) *For $\alpha > 0$, the map $(\alpha + \mathcal{L})^{-1} : L^2(G, M_{n,2}) \longrightarrow L^2(G, M_{n,2})$ is positivity preserving.*

Proof. (i) $\Rightarrow$ (ii). Let $\varphi \in D(\mathcal{L}^{1/2})$. Then by positivity preserving of T_t, we have

$$
\begin{aligned}
\langle T_t \varphi, \varphi \rangle &= \langle T_t \varphi^+ - T_t \varphi^-, \varphi^+ - \varphi^- \rangle \\
&= \langle T_t \varphi^+, \varphi^+ \rangle + \langle T_t \varphi^-, \varphi^- \rangle - \langle T_t \varphi^+, \varphi^- \rangle - \langle T_t \varphi^-, \varphi^+ \rangle \\
&\le \langle T_t |\varphi|, |\varphi| \rangle.
\end{aligned}
$$

Hence

$$
\frac{1}{t} \langle (I - T_t) |\varphi|, |\varphi| \rangle \le \frac{1}{t} \langle (I - T_t) \varphi, \varphi \rangle
$$

and $\limsup_{t \to 0} \frac{1}{t} \langle (I - T_t) |\varphi|, |\varphi| \rangle \le \langle \mathcal{L}^{1/2} \varphi, \mathcal{L}^{1/2} \varphi \rangle$. It follows that $|\varphi| \in D(\mathcal{L}^{1/2})$ and $Q(|\varphi|) \le Q(\varphi)$.

(ii) $\Leftrightarrow$ (iii). This follows from

$$
4Q(\varphi^+, \varphi^-) = Q(|\varphi|) - Q(\varphi)
$$

where $\varphi, |\varphi| \in D(\mathcal{L}^{1/2})$ implies that $\varphi^{\pm} \in D(\mathcal{L}^{1/2})$.

(iii) $\Rightarrow$ (iv). Fix $\alpha > 0$. Denote $K = D(\mathcal{L}^{1/2})$ which is a Hilbert space with respect to the inner product

$$
\langle \psi, \varphi \rangle_1 = \langle \mathcal{L}^{1/2} \psi, \mathcal{L}^{1/2} \varphi \rangle + \alpha \langle \psi, \varphi \rangle.
$$

Let $J : K \longrightarrow L^2(G, M_{n,2})$ be the natural embedding. Then, for $\psi \in K$, $f \in L^2(G, M_{n,2})$, we have

$$
\begin{aligned}
\langle \psi, (\alpha + \mathcal{L})^{-1} f \rangle_1 &= \langle \mathcal{L}^{1/2} \psi, \mathcal{L}^{1/2} (\alpha + \mathcal{L})^{-1} f \rangle + \alpha \langle \psi, (\alpha + \mathcal{L})^{-1} f \rangle \\
&= \langle (\alpha + \mathcal{L}) \psi, (\alpha + \mathcal{L})^{-1} f \rangle \\
&= \langle \psi, f \rangle = \langle J \psi, f \rangle.
\end{aligned}
$$

Therefore $J^* f = (\alpha + \mathcal{L})^{-1} f$. Let $\psi = J^* f$. We have

$$
\begin{aligned}
\langle |\psi|, |\psi| \rangle_1 &= Q(|\psi|) + \alpha \langle |\psi|, |\psi| \rangle \\
&\le Q(\psi) + \alpha \langle \psi, \psi \rangle = \langle \psi, \psi \rangle_1.
\end{aligned}
$$

Let $f \ge 0$. Then

$$
\begin{aligned}
\langle |\psi|, \psi \rangle_1 &= \langle |\psi|, J^* f \rangle_1 \\
&= \langle |\psi|, f \rangle \\
&\ge \langle \psi, f \rangle = \langle \psi, J^* f \rangle_1 = \langle \psi, \psi \rangle_1.
\end{aligned}
$$

Hence $(\alpha + \mathcal{L})^{-1} f = J^* f = \psi = |\psi| \ge 0$.

(iv) $\Rightarrow$ (i). This follows from

$$T_t = \lim_{n \to \infty} \left(I + \frac{t}{n}\mathcal{L}\right)^{-n}.$$

$\square$

Given two functions $f, h \in L^2(G, M_{n,2})$, we denote by $\langle f, h \rangle_{hs}$ the function $x \in G \mapsto \mathrm{Tr}(f(x)h(x)^*)$, and define

$$|f|_{hs} = \langle f, f \rangle_{hs}^{1/2}$$

which gives $|f|_{hs}(x) = \|f(x)\|_{hs}$ for $x \in G$.

To simplify notation, we will write $|f|$ for $|f|_{hs}$ in the rest of the chapter where confusion is unlikely.

Lemma 4.2.2. *Let $h \in L^s(G, M_{n,2}) \setminus \{0\}$ for all $s \in (1, p)$. Then $\|h\|_s$ is a differentiable function of s and*

$$\frac{d}{ds}\|h\|_s = \frac{1}{s}\|h\|_s^{1-s}\left(\int_G |h|^s \log|h| d\lambda - \|h\|_s^s \log\|h\|_s\right).$$

Proof. This follows from a simple computation of

$$\frac{d}{ds}\|h\|_s = \frac{d}{ds}\left(\int_G |h|^s\right)^{1/s}$$

$$= \|h\|_s\left(-\frac{1}{s^2}\log\int_G |h|^s + \frac{1}{s\|h\|_s^s}\int_G \frac{d}{ds}|h|^s\right)$$

where the integrals are with respect to the Haar measure λ. $\square$

In the remaining chapter, the integrals on G are with respect to the Haar measure λ.

Lemma 4.2.3. *Let $\{T_t\}_{t \geq 0}$ be a strongly continuous contractive semigroup on $L^p(G, M_{n,2})$ where $T_0 = I$ and $1 < p < \infty$. Let $c > 0$ and $p(t)$ be a real continuously differentiable function on $[0, c)$, with infimum $p(0) = p$. Given $\varphi \in C_c^\infty(G, M_{n,2})$, the function $\|T_t\varphi\|_{p(t)}$ is differentiable at $t = 0$ and*

$$\frac{d}{dt}\bigg|_{t=0} \|T_t\varphi\|_{p(t)} = \mathrm{Re}\,\|\varphi\|_p^{1-p}\int_G |\varphi|^{p-2}\left\langle \frac{d}{dt}\bigg|_{t=0} T_t\varphi, \varphi \right\rangle_{hs}$$

$$+ \frac{p'(0)\|\varphi\|_p^{1-p}}{p}\left(\int_G |\varphi|^p \log|\varphi| - \|\varphi\|_p^p \log\|\varphi\|_p\right).$$

Proof. We have

$$\frac{d}{dt}\bigg|_{t=0}\|T_t\varphi\|_{p(t)} = \lim_{t\downarrow 0}\frac{1}{t}\left(\|T_t\varphi\|_{p(t)} - \|T_0\varphi\|_p\right)$$

$$= \lim_{t\downarrow 0}\left\{\frac{1}{t}\left(\|T_t\varphi\|_{p(t)} - \|T_t\varphi\|_p\right) + \frac{1}{t}\left(\|T_t\varphi\|_p - \|\varphi\|_p\right)\right\}$$

where, by the chain rule and Proposition 2.2.5, we have

$$\lim_{t\downarrow 0}\frac{1}{t}\left(\|T_t\varphi\|_p - \|\varphi\|_p\right) = \partial\|T_0\varphi\|_p\left(\frac{d}{dt}\bigg|_{t=0}T_t\varphi\right)$$

$$= \mathrm{Re}\,\|\varphi\|_p^{1-p}\int_G|\varphi|^{p-2}\mathrm{Tr}\left(\varphi^*\left(\frac{d}{dt}\bigg|_{t=0}T_t\varphi\right)\right).$$

By the mean value theorem and Lemma 4.2.2, there exists $t_1 \in (0,t)$ such that

$$\frac{1}{t}\left(\|T_t\varphi\|_{p(t)} - \|T_t\varphi\|_p\right)$$

$$= \frac{p'(t_1)}{p(t_1)}\|T_t\varphi\|_{p(t_1)}^{1-p(t_1)}\left(\int_G|T_t\varphi|^{p(t_1)}\log|T_t\varphi| - \|T_t\varphi\|_{p(t_1)}^{p(t_1)}\log\|T_t\varphi\|_{p(t_1)}\right).$$

Letting $t \to 0$ above and putting the two limits together, we get the result.   □

Remark 4.2.4. In the above lemma, we can write

$$\mathrm{Re}\,\|\varphi\|_p^{1-p}\int_G|\varphi|^{p-2}\mathrm{Tr}\left(\varphi^*\left(\frac{d}{dt}\bigg|_{t=0}T_t\varphi\right)\right) = \frac{\|\varphi\|_p^{1-p}}{2}\int_G|\varphi|^{p-2}\frac{d}{dt}\bigg|_{t=0}|T_t\varphi|^2.$$

Indeed, by the chain rule and the Gateaux derivative of the Hilbert-Schmidt norm $\|\cdot\|_{hs}$, we have, for $x \in G$,

$$\frac{d}{dt}|T_t\varphi|^2(x) = 2|T_t\varphi|(x)\frac{d}{dt}\|T_t\varphi(x)\|_{hs}$$

$$= 2\|T_t\varphi(x)\|_{hs}\partial\|T_t\varphi(x)\|_{hs}\left(\frac{d}{dt}T_t\varphi(x)\right)$$

$$= 2\mathrm{Re}\,\mathrm{Tr}(T_t\varphi(x)^*T_t{}'\varphi(x)).$$

The following result for matrix semigroups, extending a result of Gross in [36], answers the above question of smoothing the semigroup $\{T_t\}_{t\geq 0}$ induced by a convolution semigroup $\{\sigma_t\}_{t>0}$ of M_n-valued measures satisfying $\widetilde{\sigma}_t^* = \sigma_t \geq 0$. The condition for hypercontractivity of $\{T_t\}_{t\geq 0}$ is a log-Sobolev type inequality of index p for its generator $\mathcal{L}$, for each $p \in (1,\infty)$. The proof is based on a differential inequality, as in [36]. See also [3, 17].

We say that an operator $\mathcal{L}$ in $L^2(G,M_{n,2})$ generates a *contractive* semigroup $\{T_t\}_{t\geq 0}$ on $L^p(G,M_{n,2})$ if each T_t maps $L^2(G,M_{n,2}) \cap L^p(G,M_{n,2})$ into $L^2(G,M_{n,2})) \cap L^p(G,M_{n,2})$, and is contractive in the L^p-norm.

Theorem 4.2.5. *Let* $-\mathcal{L}$ *be a self-adjoint operator in* $L^2(G, M_{n,2})$, *generating a positive strongly continuous contractive semigroup* $\{T_t\}_{t\geq 0}$ *on* $L^p(G, M_{n,2})$ *for all* p. *Given* $a > 0$ *and* $b \geq 0$, *the following conditions are equivalent.*

(i) $\{T_t\}_{t\geq 0}$ *is hypercontractive, that is, for each* $p \in (1, \infty)$ *and* $t > 0$,

$$\|T_t\varphi\|_{p(t)} \leq e^{m(t)}\|\varphi\|_p \qquad (\varphi \in C_c^\infty(G, M_n))$$

where $\quad p(t) = 1 + (p-1)e^{4t/a}$ *and* $m(t) = b\left(p^{-1} - p(t)^{-1}\right)$.

(ii) *For each* $p \in (1, \infty)$ *and* $\varphi \in C_c(G, M_{n,2})$, $\mathcal{L}$ *satisfies the inequality*

$$\int_G |\varphi|^p \log|\varphi|^p d\lambda \leq -\frac{ap^2}{4(p-1)}\int_G |\varphi|^{p-2}\mathrm{Re}\,\langle\mathcal{L}\varphi, \varphi\rangle_{hs}d\lambda + \|\varphi\|_p^p(b + \log\|\varphi\|_p^p).$$

Proof. Let $\varphi \in C_c^\infty(G, M_{n,2})\backslash\{0\}$. For $t \in [0, \infty)$, let $F(t) = e^{-m(t)}\|T_t\varphi\|_{p(t)}$ where $m(0) = 0$ and $p(0) = p$. We have

$$\frac{d}{dt}\log F(t) = -m'(t) + \frac{1}{\|T_t\varphi\|_{p(t)}}\frac{d}{dt}\|T_t\varphi\|_{p(t)}$$

where, by Lemma 4.2.3, we have

$$\frac{d}{dt}\bigg|_{t=0}\log F(t) = -m'(0) + \frac{1}{\|\varphi\|_p^p}\int_G |\varphi|^{p-2}\mathrm{Re}\,\langle\mathcal{L}\varphi, \varphi\rangle_{hs}$$

$$+ \frac{p'(0)}{p^2\|\varphi\|_p^p}\left(\int_G |\varphi|^p\log|\varphi|^p - \|\varphi\|_p^p\log\|\varphi\|_p^p\right).$$

For (i) $\Rightarrow$ (ii), we note that $F'(0) \leq 0$ since $F(t) \leq F(0)$ for all t. Hence $\frac{d}{dt}\bigg|_{t=0}\log F(t) \leq 0$ and we have

$$\int_G |\varphi|^p\log|\varphi|^p \leq -\frac{p^2}{p'(0)}\int_G |\varphi|^{p-2}\mathrm{Re}\,\langle\mathcal{L}\varphi, \varphi\rangle_{hs} + \|\varphi\|_p^p\left(\frac{m'(0)p^2}{p'(0)} + \log\|\varphi\|_p^p\right).$$

We note that $p(t)$ and $m(t)$ solve the differential equations

$$\frac{p(t)^2}{p'(t)} = \frac{ap^2}{4(p-1)}, \quad p(0) = p$$

and

$$\frac{m'(t)p(t)^2}{p'(t)} = b, \quad m(0) = 0.$$

Hence we have

$$\frac{p^2}{p'(0)} = \frac{ap^2}{4(p-1)} \quad \text{and} \quad \frac{m'(0)p^2}{p'(0)} = b$$

and (ii) holds.

Conversely, (ii) implies

$$\frac{F'(0)}{F(0)} = \frac{d}{dt}\bigg|_{t=0} \log F(t) \le 0$$

where $F(0) = \|\varphi\|_p$. It follows that $F(t) \le F(0)$ for all t, giving

$$e^{-m(t)}\|T_t\varphi\|_{p(t)} \le \|\varphi\|_p.$$

$\square$

Remark 4.2.6. In the scalar case $\varphi \in C_c^\infty(G)$, we have

$$\int_G |\varphi|^{p-2}\mathrm{Re}\,\langle \mathcal{L}\varphi, \varphi\rangle_{hs} = \mathrm{Re}\int_G \mathcal{L}\varphi|\varphi|^{p-2}\overline{\varphi}$$

which can be written as $\mathrm{Re}\int_G \mathcal{L}\varphi\overline{\varphi_p}$ where $\varphi_p = (\mathrm{sgn}\,\varphi)|\varphi|^{p-1}$ and

$$\mathrm{sgn}\,z = \begin{cases} \frac{z}{|z|} & \text{if } z \ne 0, \\ 0 & \text{if } z = 0. \end{cases}$$

Hence, modulo some constants, the inequality in Theorem 4.2.5 (ii) is identical to that in (2.1) of [36].

References

1. R. Azencott, Espaces de Poisson des groupes localement compacts, Lecture Notes in Math. **148**, Springer-Verlag, Berlin, 1970.
2. B. Blackadar, K-theory for operator algebras, MSRI Publ. **5**, Springer-Verlag, Heidelberg, 1986.
3. D. Bakry, *On Sobolev and logarithmic Sobolev inequalities for Markov semigroups*, in "New Trends in Stochastic Analysis" (ed. K. D. Elworthy) World Scientific (1997) 43–75.
4. C. Berg and J.P.R. Christensen, *On the relation between amenability of locally compact groups and the norms of convolution operators*, Math. Ann. **208** (1974) 149–153.
5. J.F. Bonnans and A. Shapiro, Perturbation analysis of optimization problems, Springer-Verlag, Berlin, 2000.
6. A. Böttcher, Y.I. Karlovich and I.M. Spitkovsky, Convolution operators and factorization of almost periodic matrix functions, Birkhäuser Verlag, Berlin, 2002.
7. C. Chen and C-H. Chu, *Spectrum of a homogeneous graph*, J. Math. Anal. Appl. (to appear).
8. G. Choquet and J. Deny, *Sur l'équation de convolution $\mu = \mu * \sigma$*, C.R. Acad. Sc. Paris **250** (1960) 779–801.
9. C-H. Chu, *Matrix-valued harmonic functions on groups*, J. Reine Angew. Math. **552** (2002) 15–52.
10. C-H. Chu, *Harmonic function spaces on groups*, J. London Math. Soc. **70** (2004) 182–198.
11. C-H. Chu, *Jordan triples and Riemannian symmetric spaces*, Preprint (2008).
12. C-H. Chu, T. Hilberdink and J. Howroyd, *A matrix-valued Choquet-Deny Theorem*, Proc. Amer. Math. Soc. **129** (2001) 229–235.
13. C-H. Chu and A.T.M. Lau, *Harmonic functions on groups and Fourier algebras*, Lecture Notes in Math. **1782**, Springer-Verlag, Heidelberg, 2002.
14. C-H. Chu and A.T-M. Lau, *Jordan structures in harmonic functions and Fourier algebras on homogeneous spaces*, Math. Ann. **336** (2006) 803–840.
15. C-H. Chu and C-W. Leung, *The convolution equation of Choquet and Deny on* [IN]*-groups*, Integr. Equat. Oper. Th. **40** (2001) 391–402.
16. C-H. Chu and T.G. Vu, *A Liouville theorem for matrix-valued harmonic functions on nilpotent groups*, Bull. London Math. Soc. **35** (2003) 651–658.
17. C-H. Chu and Z. Qian, *Dirichlet forms and Markov semigroups on non-associative vector bundles*, Studies in Adv. Math., Amer. Math. Soc. (to appear).
18. C-H. Chu and N-C. Wong, *Isometries between C*-algebras*, Rev. Mat. Iberoamericana **20** (2004) 87–105.
19. F.R.K. Chung, *Spectral graph theory*, CMBS Leture Notes, Amer. Math. Sco. Providence, 1997.
20. F.R.K. Chung and S. Sternberg, *Laplacian and vibrational spectra for homogeneous graphs*, J. Graph Theory **16** (1992) 605–627.

21. E.B. Davies and O.S. Rothaus, *Markov semigroups on C* bundles*, J. Funct. Analysis **85** (1989) 264–286.
22. Diestel and Uhl, Vector measures, (Math. Surveys **15**) Amer. Math. Soc. Providence, 1977.
23. J. Dixmier, Les C*-algèbres et leur représentations, Gauthier-Villar, Paris, 1969.
24. N. Dunford and J.T. Schwartz, Linear Operators I, J. Wiley & Sons, New York, 1988.
25. E.B. Dynkin, Markov Process, Vol. I, Springer, Berlin, 1965.
26. R. J. Elliot, *Two notes on spectral synthesis for discrete abelian groups*, Proc. Camb. Phil. Soc. **61** (1965) 617–620.
27. E.B. Folland, A course in abstract harmonic analysis, CRC Press, Boca Raton, 1995.
28. Y. Friedman and B. Russo, *Solution of the contractive projection problem*, J. Funct. Analysis **60** (1985) 56–79.
29. H. Furstenberg, *A Poisson formula for semi-simple Lie groups*, Ann. of Math. **77** (1963) 335–386.
30. H. Furstenberg, *Boundaries of Riemannian symmetric spaces*, in ('*Symmetric spaces*', Pure Appl. Math. 8, Marcel Dekker, 1972) 359–377.
31. J. E.Gilbert, *Specral synthesis problems for invariant subspaces on groups*, Amer. J. Math. **88** (1966) 626–635.
32. R.E. Greene and H. Wu, 'Integrals of subharmonic functions on manifolds of nonnegative curvature', Invent. Math. **27** (1974) 265–298.
33. F.P. Greenleaf, Invariant means on topological groups, van Nostrand, New York, 1969.
34. N.E. Gretsky and J.J. Uhl, *Bounded linear operators on Banach function spaces of vector-valued functions*, Trans. Amer. Math. Soc. **167** (1972) 263–277.
35. R.I. Grigorchuk, P. Linnell, T. Schick and A. Zuk, *On a question of Atiyah*, C.R. Acad. Sci. Paris, t. **331**, Série I (2000) 663–668.
36. L. Gross, *Logarithmic Sobolev inequalities*, Amer. J. Math. **97** (1975) 1061–1083.
37. S. Helgason, Differential geometry, Lie groups and symmetric spaces, Academic Press 1980.
38. E. Hewitt and K.A. Ross, Abstract harmonic analysis, Vol. I, Springer-Verlag, Berlin, 1963.
39. G.A. Hunt, *Semi-groups of measures on Lie groups*, Trans. Amer. Math. Soc. **81** (1956), 264–293.
40. B.E. Johnson, *Harmonic functions on nilpotent groups*, Integral Equations & Oper. Th. **40** (2001) 454–464.
41. W. Kaup, *A Riemann mapping theorem for bounded symmetric domains in complex Banach spaces*, Math. Z. **183** (1983) 503–529.
42. W. Kaup, *Contractive projections on Jordan C*-algebras and generalizations*, Math. Scand. **54** (1984) 95–100.
43. M. Koecher, *Jordan algebras and differential geometry*, Proc. ICM (Nice 1970) 279–283.
44. R. Larsen, An introduction to the theory of multipliers, Springer-Verlag, Berlin, 1971.
45. K-S. Lau, J. Wang and C-H. Chu, *Vector-valued Choquet–Deny theorem, renewal equation and self-similar measures*, Studia Math. **117** (1995) 1–28.
46. P. Li and R. Schoen, 'L^p and mean value properties of subharmonic functions on Riemannian manifolds', Acta Math. **153** (1984) 279–301.
47. O. Loos, Bounded symmetric domains and Jordan pairs (Mathematical Lectures) University of California, Irvine 1977.
48. F. Lust-Piquard and W. Schachermayer, *Functions in $L^\infty(G)$ and associated convolution operators*, Studia Math. **93** (1989) 109–136.
49. B. Malgrange, *Existence et approximation des solutions des équations aux dérivées partielles et des équations de convolution*, Ann. l'Institut Fourier **6** (1955/56) 271–355.
50. J. Milnor, *Curvatures of left invariant metrics on Lie groups*, Adv. Math. 21 (1976) 293–329.
51. G.K. Pedersen, C*-algebras and their automorphism groups, Academic Press, London, 1979.
52. F. Parreau, *Measures with real spectra*, Invent. Math. **98** (1989) 311–330.
53. R.R. Phelps, Convex functions, monotone operators and differentiability, Lecture Notes in Math. **1364**, Springer-Verlag, Heidelberg, 1989.
54. V. Rabinovich, S. Roch and B. Silbermann, Limit operators and their applications in operator theory, Birkhäuser Verlag, Basel, 2004.

55. W. Rudin, Fourier analysis on groups, Interscience Publishers, New York, 1962.
56. P. Sarnak, *Spectra of singular measures as mulitpliers on L^p*, J. Funct. Analysis **37** (1980) 302–317.
57. I. Satake, Algebraic structures of symmetric domains, Princeton Univ. Press, Princeton, 1980.
58. L. Schwartz, *Théorie génerale des fonctions moyennes-périodiques*, Ann. of Math. **48** (1947) 857–929.
59. B. Ya Shteinberg, *Convolution type operators on locally compact groups*, Funktsional. Anal. i Prilozhen. **15** (1981) 95–96.
60. B. Ya Shteinberg, *Boundedness and compactness of convolution operators with unbounded coefficients on locally compact groups*, Mat. Zametki **38** (1985) 278–292.
61. B. Ya Shteinberg, *Compactification of locally compact groups and Fredholmness of convolution operators with coefficients in factor groups*, Tr. St.-Peterbg. Mat. Obshch. **6** (1998) 242–260.
62. M. Takesaki, Theory of operator algebras I, Springer-Verlag, Berlin, 1979.
63. H. Upmeier, Symmetric Banach manifolds and Jordan C*-algebras (North Holland Math. Studies **104**) North Holland, Amsterdam, 1985.
64. S-T. Yau, *Some function-theoretic properties of complete Riemannian manifolds and their applications to geometry*, Indiana Univ. Math. J. **25** (1976) 659–670.
65. S-T. Yau, *Harmonic functions on complete Riemannian manifolds*, Comm. Pure Applied Math. **28** (1975) 201–228.
66. M. Zafran, *The spectra of multiplier transformations on the L_p spaces*, Ann. of Math. **103** (1976) 355–374.

List of Symbols

Index

Lecture Notes in Mathematics

For information about earlier volumes
please contact your bookseller or Springer
LNM Online archive: springerlink.com

Vol. 1817: E. Koelink, W. Van Assche (Eds.), Orthogonal Polynomials and Special Functions. Leuven 2002 (2003)

Vol. 1818: M. Bildhauer, Convex Variational Problems with Linear, nearly Linear and/or Anisotropic Growth Conditions (2003)

Vol. 1819: D. Masser, Yu. V. Nesterenko, H. P. Schlickewei, W. M. Schmidt, M. Waldschmidt, Diophantine Approximation. Cetraro, Italy 2000. Editors: F. Amoroso, U. Zannier (2003)

Vol. 1820: F. Hiai, H. Kosaki, Means of Hilbert Space Operators (2003)

Vol. 1821: S. Teufel, Adiabatic Perturbation Theory in Quantum Dynamics (2003)

Vol. 1822: S.-N. Chow, R. Conti, R. Johnson, J. Mallet-Paret, R. Nussbaum, Dynamical Systems. Cetraro, Italy 2000. Editors: J. W. Macki, P. Zecca (2003)

Vol. 1823: A. M. Anile, W. Allegretto, C. Ringhofer, Mathematical Problems in Semiconductor Physics. Cetraro, Italy 1998. Editor: A. M. Anile (2003)

Vol. 1824: J. A. Navarro González, J. B. Sancho de Salas, $\mathscr{C}^{\infty}$ – Differentiable Spaces (2003)

Vol. 1825: J. H. Bramble, A. Cohen, W. Dahmen, Multiscale Problems and Methods in Numerical Simulations, Martina Franca, Italy 2001. Editor: C. Canuto (2003)

Vol. 1826: K. Dohmen, Improved Bonferroni Inequalities via Abstract Tubes. Inequalities and Identities of Inclusion-Exclusion Type. VIII, 113 p, 2003.

Vol. 1827: K. M. Pilgrim, Combinations of Complex Dynamical Systems. IX, 118 p, 2003.

Vol. 1828: D. J. Green, Gröbner Bases and the Computation of Group Cohomology. XII, 138 p, 2003.

Vol. 1829: E. Altman, B. Gaujal, A. Hordijk, Discrete-Event Control of Stochastic Networks: Multimodularity and Regularity. XIV, 313 p, 2003.

Vol. 1830: M. I. Gil', Operator Functions and Localization of Spectra. XIV, 256 p, 2003.

Vol. 1831: A. Connes, J. Cuntz, E. Guentner, N. Higson, J. E. Kaminker, Noncommutative Geometry, Martina Franca, Italy 2002. Editors: S. Doplicher, L. Longo (2004)

Vol. 1832: J. Azéma, M. Émery, M. Ledoux, M. Yor (Eds.), Séminaire de Probabilités XXXVII (2003)

Vol. 1833: D.-Q. Jiang, M. Qian, M.-P. Qian, Mathematical Theory of Nonequilibrium Steady States. On the Frontier of Probability and Dynamical Systems. IX, 280 p, 2004.

Vol. 1834: Yo. Yomdin, G. Comte, Tame Geometry with Application in Smooth Analysis. VIII, 186 p, 2004.

Vol. 1835: O.T. Izhboldin, B. Kahn, N.A. Karpenko, A. Vishik, Geometric Methods in the Algebraic Theory of Quadratic Forms. Summer School, Lens, 2000. Editor: J.-P. Tignol (2004)

Vol. 1836: C. Năstăsescu, F. Van Oystaeyen, Methods of Graded Rings. XIII, 304 p, 2004.

Vol. 1837: S. Tavaré, O. Zeitouni, Lectures on Probability Theory and Statistics. Ecole d'Eté de Probabilités de Saint-Flour XXXI-2001. Editor: J. Picard (2004)

Vol. 1838: A.J. Ganesh, N.W. O'Connell, D.J. Wischik, Big Queues. XII, 254 p, 2004.

Vol. 1839: R. Gohm, Noncommutative Stationary Processes. VIII, 170 p, 2004.

Vol. 1840: B. Tsirelson, W. Werner, Lectures on Probability Theory and Statistics. Ecole d'Eté de Probabilités de Saint-Flour XXXII-2002. Editor: J. Picard (2004)

Vol. 1841: W. Reichel, Uniqueness Theorems for Variational Problems by the Method of Transformation Groups (2004)

Vol. 1842: T. Johnsen, A. L. Knutsen, K3 Projective Models in Scrolls (2004)

Vol. 1843: B. Jefferies, Spectral Properties of Noncommuting Operators (2004)

Vol. 1844: K.F. Siburg, The Principle of Least Action in Geometry and Dynamics (2004)

Vol. 1845: Min Ho Lee, Mixed Automorphic Forms, Torus Bundles, and Jacobi Forms (2004)

Vol. 1846: H. Ammari, H. Kang, Reconstruction of Small Inhomogeneities from Boundary Measurements (2004)

Vol. 1847: T.R. Bielecki, T. Björk, M. Jeanblanc, M. Rutkowski, J.A. Scheinkman, W. Xiong, Paris-Princeton Lectures on Mathematical Finance 2003 (2004)

Vol. 1848: M. Abate, J. E. Fornaess, X. Huang, J. P. Rosay, A. Tumanov, Real Methods in Complex and CR Geometry, Martina Franca, Italy 2002. Editors: D. Zaitsev, G. Zampieri (2004)

Vol. 1849: Martin L. Brown, Heegner Modules and Elliptic Curves (2004)

Vol. 1850: V. D. Milman, G. Schechtman (Eds.), Geometric Aspects of Functional Analysis. Israel Seminar 2002-2003 (2004)

Vol. 1851: O. Catoni, Statistical Learning Theory and Stochastic Optimization (2004)

Vol. 1852: A.S. Kechris, B.D. Miller, Topics in Orbit Equivalence (2004)

Vol. 1853: Ch. Favre, M. Jonsson, The Valuative Tree (2004)

Vol. 1854: O. Saeki, Topology of Singular Fibers of Differential Maps (2004)

Vol. 1855: G. Da Prato, P.C. Kunstmann, I. Lasiecka, A. Lunardi, R. Schnaubelt, L. Weis, Functional Analytic Methods for Evolution Equations. Editors: M. Iannelli, R. Nagel, S. Piazzera (2004)

Vol. 1856: K. Back, T.R. Bielecki, C. Hipp, S. Peng, W. Schachermayer, Stochastic Methods in Finance, Bressanone/Brixen, Italy, 2003. Editors: M. Fritelli, W. Runggaldier (2004)

Vol. 1857: M. Émery, M. Ledoux, M. Yor (Eds.), Séminaire de Probabilités XXXVIII (2005)

Vol. 1858: A.S. Cherny, H.-J. Engelbert, Singular Stochastic Differential Equations (2005)

Vol. 1859: E. Letellier, Fourier Transforms of Invariant Functions on Finite Reductive Lie Algebras (2005)

Vol. 1860: A. Borisyuk, G.B. Ermentrout, A. Friedman, D. Terman, Tutorials in Mathematical Biosciences I. Mathematical Neurosciences (2005)

Vol. 1861: G. Benettin, J. Henrard, S. Kuksin, Hamiltonian Dynamics – Theory and Applications, Cetraro, Italy, 1999. Editor: A. Giorgilli (2005)

Vol. 1862: B. Helffer, F. Nier, Hypoelliptic Estimates and Spectral Theory for Fokker-Planck Operators and Witten Laplacians (2005)

Vol. 1863: H. Führ, Abstract Harmonic Analysis of Continuous Wavelet Transforms (2005)

Vol. 1864: K. Efstathiou, Metamorphoses of Hamiltonian Systems with Symmetries (2005)

Vol. 1865: D. Applebaum, B.V. R. Bhat, J. Kustermans, J. M. Lindsay, Quantum Independent Increment Processes I. From Classical Probability to Quantum Stochastic Calculus. Editors: M. Schürmann, U. Franz (2005)

Vol. 1866: O.E. Barndorff-Nielsen, U. Franz, R. Gohm, B. Kümmerer, S. Thorbjønsen, Quantum Independent Increment Processes II. Structure of Quantum Lévy Processes, Classical Probability, and Physics. Editors: M. Schürmann, U. Franz, (2005)

Vol. 1867: J. Sneyd (Ed.), Tutorials in Mathematical Biosciences II. Mathematical Modeling of Calcium Dynamics and Signal Transduction. (2005)

Vol. 1868: J. Jorgenson, S. Lang, Pos$_n$(R) and Eisenstein Series. (2005)

Vol. 1869: A. Dembo, T. Funaki, Lectures on Probability Theory and Statistics. Ecole d'Eté de Probabilités de Saint-Flour XXXIII-2003. Editor: J. Picard (2005)

Vol. 1870: V.I. Gurariy, W. Lusky, Geometry of Müntz Spaces and Related Questions. (2005)

Vol. 1871: P. Constantin, G. Gallavotti, A.V. Kazhikhov, Y. Meyer, S. Ukai, Mathematical Foundation of Turbulent Viscous Flows, Martina Franca, Italy, 2003. Editors: M. Cannone, T. Miyakawa (2006)

Vol. 1872: A. Friedman (Ed.), Tutorials in Mathematical Biosciences III. Cell Cycle, Proliferation, and Cancer (2006)

Vol. 1873: R. Mansuy, M. Yor, Random Times and Enlargements of Filtrations in a Brownian Setting (2006)

Vol. 1874: M. Yor, M. Émery (Eds.), In Memoriam Paul-André Meyer - Séminaire de Probabilités XXXIX (2006)

Vol. 1875: J. Pitman, Combinatorial Stochastic Processes. Ecole d'Eté de Probabilités de Saint-Flour XXXII-2002. Editor: J. Picard (2006)

Vol. 1876: H. Herrlich, Axiom of Choice (2006)

Vol. 1877: J. Steuding, Value Distributions of L-Functions (2007)

Vol. 1878: R. Cerf, The Wulff Crystal in Ising and Percolation Models, Ecole d'Eté de Probabilités de Saint-Flour XXXIV-2004. Editor: Jean Picard (2006)

Vol. 1879: G. Slade, The Lace Expansion and its Applications, Ecole d'Eté de Probabilités de Saint-Flour XXXIV-2004. Editor: Jean Picard (2006)

Vol. 1880: S. Attal, A. Joye, C.-A. Pillet, Open Quantum Systems I, The Hamiltonian Approach (2006)

Vol. 1881: S. Attal, A. Joye, C.-A. Pillet, Open Quantum Systems II, The Markovian Approach (2006)

Vol. 1882: S. Attal, A. Joye, C.-A. Pillet, Open Quantum Systems III, Recent Developments (2006)

Vol. 1883: W. Van Assche, F. Marcellàn (Eds.), Orthogonal Polynomials and Special Functions, Computation and Application (2006)

Vol. 1884: N. Hayashi, E.I. Kaikina, P.I. Naumkin, I.A. Shishmarev, Asymptotics for Dissipative Nonlinear Equations (2006)

Vol. 1885: A. Telcs, The Art of Random Walks (2006)

Vol. 1886: S. Takamura, Splitting Deformations of Degenerations of Complex Curves (2006)

Vol. 1887: K. Habermann, L. Habermann, Introduction to Symplectic Dirac Operators (2006)

Vol. 1888: J. van der Hoeven, Transseries and Real Differential Algebra (2006)

Vol. 1889: G. Osipenko, Dynamical Systems, Graphs, and Algorithms (2006)

Vol. 1890: M. Bunge, J. Funk, Singular Coverings of Toposes (2006)

Vol. 1891: J.B. Friedlander, D.R. Heath-Brown, H. Iwaniec, J. Kaczorowski, Analytic Number Theory, Cetraro, Italy, 2002. Editors: A. Perelli, C. Viola (2006)

Vol 1892: A Baddeley, I. Bárány, R. Schneider, W. Weil, Stochastic Geometry, Martina Franca, Italy, 2004. Editor: W. Weil (2007)

Vol. 1893: H. Hanßmann, Local and Semi-Local Bifurcations in Hamiltonian Dynamical Systems, Results and Examples (2007)

Vol. 1894: C.W. Groetsch, Stable Approximate Evaluation of Unbounded Operators (2007)

Vol. 1895: L. Molnár, Selected Preserver Problems on Algebraic Structures of Linear Operators and on Function Spaces (2007)

Vol. 1896: P. Massart, Concentration Inequalities and Model Selection, Ecole d'Été de Probabilités de Saint-Flour XXXIII-2003. Editor: J. Picard (2007)

Vol. 1897: R. Doney, Fluctuation Theory for Lévy Processes, Ecole d'Été de Probabilités de Saint-Flour XXXV-2005. Editor: J. Picard (2007)

Vol. 1898: H.R. Beyer, Beyond Partial Differential Equations, On linear and Quasi-Linear Abstract Hyperbolic Evolution Equations (2007)

Vol. 1899: Séminaire de Probabilités XL. Editors: C. Donati-Martin, M. Émery, A. Rouault, C. Stricker (2007)

Vol. 1900: E. Bolthausen, A. Bovier (Eds.), Spin Glasses (2007)

Vol. 1901: O. Wittenberg, Intersections de deux quadriques et pinceaux de courbes de genre 1, Intersections of Two Quadrics and Pencils of Curves of Genus 1 (2007)

Vol. 1902: A. Isaev, Lectures on the Automorphism Groups of Kobayashi-Hyperbolic Manifolds (2007)

Vol. 1903: G. Kresin, V. Maz'ya, Sharp Real-Part Theorems (2007)

Vol. 1904: P. Giesl, Construction of Global Lyapunov Functions Using Radial Basis Functions (2007)

Vol. 1905: C. Prévôt, M. Röckner, A Concise Course on Stochastic Partial Differential Equations (2007)

Vol. 1906: T. Schuster, The Method of Approximate Inverse: Theory and Applications (2007)

Vol. 1907: M. Rasmussen, Attractivity and Bifurcation for Nonautonomous Dynamical Systems (2007)

Vol. 1908: T.J. Lyons, M. Caruana, T. Lévy, Differential Equations Driven by Rough Paths, Ecole d'Été de Probabilités de Saint-Flour XXXIV-2004 (2007)

Vol. 1909: H. Akiyoshi, M. Sakuma, M. Wada, Y. Yamashita, Punctured Torus Groups and 2-Bridge Knot Groups (I) (2007)

Vol. 1910: V.D. Milman, G. Schechtman (Eds.), Geometric Aspects of Functional Analysis. Israel Seminar 2004-2005 (2007)

Vol. 1911: A. Bressan, D. Serre, M. Williams, K. Zumbrun, Hyperbolic Systems of Balance Laws. Cetraro, Italy 2003. Editor: P. Marcati (2007)

Vol. 1912: V. Berinde, Iterative Approximation of Fixed Points (2007)

Vol. 1913: J.E. Marsden, G. Misiołek, J.-P. Ortega, M. Perlmutter, T.S. Ratiu, Hamiltonian Reduction by Stages (2007)

Vol. 1914: G. Kutyniok, Affine Density in Wavelet Analysis (2007)

Vol. 1915: T. Bıyıkoğlu, J. Leydold, P.F. Stadler, Laplacian Eigenvectors of Graphs. Perron-Frobenius and Faber-Krahn Type Theorems (2007)

Vol. 1916: C. Villani, F. Rezakhanlou, Entropy Methods for the Boltzmann Equation. Editors: F. Golse, S. Olla (2008)

Vol. 1917: I. Veselić, Existence and Regularity Properties of the Integrated Density of States of Random Schrödinger (2008)

Vol. 1918: B. Roberts, R. Schmidt, Local Newforms for GSp(4) (2007)

Vol. 1919: R.A. Carmona, I. Ekeland, A. Kohatsu-Higa, J.-M. Lasry, P.-L. Lions, H. Pham, E. Taflin, Paris-Princeton Lectures on Mathematical Finance 2004.

Editors: R.A. Carmona, E. Çinlar, I. Ekeland, E. Jouini, J.A. Scheinkman, N. Touzi (2007)

Vol. 1920: S.N. Evans, Probability and Real Trees. Ecole d'Été de Probabilités de Saint-Flour XXXV-2005 (2008)

Vol. 1921: J.P. Tian, Evolution Algebras and their Applications (2008)

Vol. 1922: A. Friedman (Ed.), Tutorials in Mathematical BioSciences IV. Evolution and Ecology (2008)

Vol. 1923: J.P.N. Bishwal, Parameter Estimation in Stochastic Differential Equations (2008)

Vol. 1924: M. Wilson, Littlewood-Paley Theory and Exponential-Square Integrability (2008)

Vol. 1925: M. du Sautoy, L. Woodward, Zeta Functions of Groups and Rings (2008)

Vol. 1926: L. Barreira, V. Claudia, Stability of Nonautonomous Differential Equations (2008)

Vol. 1927: L. Ambrosio, L. Caffarelli, M.G. Crandall, L.C. Evans, N. Fusco, Calculus of Variations and Non-Linear Partial Differential Equations. Cetraro, Italy 2005. Editors: B. Dacorogna, P. Marcellini (2008)

Vol. 1928: J. Jonsson, Simplicial Complexes of Graphs (2008)

Vol. 1929: Y. Mishura, Stochastic Calculus for Fractional Brownian Motion and Related Processes (2008)

Vol. 1930: J.M. Urbano, The Method of Intrinsic Scaling. A Systematic Approach to Regularity for Degenerate and Singular PDEs (2008)

Vol. 1931: M. Cowling, E. Frenkel, M. Kashiwara, A. Valette, D.A. Vogan, Jr., N.R. Wallach, Representation Theory and Complex Analysis. Venice, Italy 2004. Editors: E.C. Tarabusi, A. D'Agnolo, M. Picardello (2008)

Vol. 1932: A.A. Agrachev, A.S. Morse, E.D. Sontag, H.J. Sussmann, V.I. Utkin, Nonlinear and Optimal Control Theory. Cetraro, Italy 2004. Editors: P. Nistri, G. Stefani (2008)

Vol. 1933: M. Petkovic, Point Estimation of Root Finding Methods (2008)

Vol. 1934: C. Donati-Martin, M. Émery, A. Rouault, C. Stricker (Eds.), Séminaire de Probabilités XLI (2008)

Vol. 1935: A. Unterberger, Alternative Pseudodifferential Analysis (2008)

Vol. 1936: P. Magal, S. Ruan (Eds.), Structured Population Models in Biology and Epidemiology (2008)

Vol. 1937: G. Capriz, P. Giovine, P.M. Mariano (Eds.), Mathematical Models of Granular Matter (2008)

Vol. 1938: D. Auroux, F. Catanese, M. Manetti, P. Seidel, B. Siebert, I. Smith, G. Tian, Symplectic 4-Manifolds and Algebraic Surfaces. Cetraro, Italy 2003. Editors: F. Catanese, G. Tian (2008)

Vol. 1939: D. Boffi, F. Brezzi, L. Demkowicz, R.G. Durán, R.S. Falk, M. Fortin, Mixed Finite Elements, Compatibility Conditions, and Applications. Cetraro, Italy 2006. Editors: D. Boffi, L. Gastaldi (2008)

Vol. 1940: J. Banasiak, V. Capasso, M.A.J. Chaplain, M. Lachowicz, J. Miękisz, Multiscale Problems in the Life Sciences. From Microscopic to Macroscopic. Będlewo, Poland 2006. Editors: V. Capasso, M. Lachowicz (2008)

Vol. 1941: S.M.J. Haran, Arithmetical Investigations. Representation Theory, Orthogonal Polynomials, and Quantum Interpolations (2008)

Vol. 1942: S. Albeverio, F. Flandoli, Y.G. Sinai, SPDE in Hydrodynamic. Recent Progress and Prospects. Cetraro, Italy 2005. Editors: G. Da Prato, M. Röckner (2008)

Vol. 1943: L.L. Bonilla (Ed.), Inverse Problems and Imaging. Martina Franca, Italy 2002 (2008)

Vol. 1944: A. Di Bartolo, G. Falcone, P. Plaumann, K. Strambach, Algebraic Groups and Lie Groups with Few Factors (2008)

Vol. 1945: F. Brauer, P. van den Driessche, J. Wu (Eds.), Mathematical Epidemiology (2008)

Vol. 1946: G. Allaire, A. Arnold, P. Degond, T.Y. Hou, Quantum Transport. Modelling, Analysis and Asymptotics. Cetraro, Italy 2006. Editors: N.B. Abdallah, G. Frosali (2008)

Vol. 1947: D. Abramovich, M. Mariño, M. Thaddeus, R. Vakil, Enumerative Invariants in Algebraic Geometry and String Theory. Cetraro, Italy 2005. Editors: K. Behrend, M. Manetti (2008)

Vol. 1948: F. Cao, J-L. Lisani, J-M. Morel, P. Musé, F. Sur, A Theory of Shape Identification (2008)

Vol. 1949: H.G. Feichtinger, B. Helffer, M.P. Lamoureux, N. Lerner, J. Toft, Pseudo-Differential Operators. Quantization and Signals. Cetraro, Italy 2006. Editors: L. Rodino, M.W. Wong (2008)

Vol. 1950: M. Bramson, Stability of Queueing Networks, Ecole d'Eté de Probabilités de Saint-Flour XXXVI-2006 (2008)

Vol. 1951: A. Moltó, J. Orihuela, S. Troyanski, M. Valdivia, A Non Linear Transfer Technique for Renorming (2008)

Vol. 1952: R. Mikhailov, I.B.S. Passi, Lower Central and Dimension Series of Groups (2008)

Vol. 1953: K. Arwini, C.T.J. Dodson, Information Geometry (2008)

Vol. 1954: P. Biane, L. Bouten, F. Cipriani, N. Konno, N. Privault, Q. Xu, Quantum Potential Theory. Editors: U. Franz, M. Schuermann (2008)

Vol. 1955: M. Bernot, V. Caselles, J.-M. Morel, Optimal transportation networks (2008)

Vol. 1956: C.-H. Chu, Matrix Convolution Operators on Groups (2008)

Vol. 1957: A. Guionnet, On Random Matrices: Macroscopic Asymptotics, Ecole d'Eté de Probabilités de Saint-Flour XXXVI-2006 (2008)

Vol. 1958: M.C. Olsson, Compactifying Moduli Spaces for Abelian Varieties (2008)

Recent Reprints and New Editions

Vol. 1702: J. Ma, J. Yong, Forward-Backward Stochastic Differential Equations and their Applications. 1999 – Corr. 3rd printing (2007)

Vol. 830: J.A. Green, Polynomial Representations of GL_n, with an Appendix on Schensted Correspondence and Littelmann Paths by K. Erdmann, J.A. Green and M. Schoker 1980 – 2nd corr. and augmented edition (2007)

Vol. 1693: S. Simons, From Hahn-Banach to Monotonicity (Minimax and Monotonicity 1998) – 2nd exp. edition (2008)

Vol. 470: R.E. Bowen, Equilibrium States and the Ergodic Theory of Anosov Diffeomorphisms. With a preface by D. Ruelle. Edited by J.-R. Chazottes. 1975 – 2nd rev. edition (2008)

Vol. 523: S.A. Albeverio, R.J. Høegh-Krohn, S. Mazzucchi, Mathematical Theory of Feynman Path Integral. 1976 – 2nd corr. and enlarged edition (2008)

Vol. 1764: A. Cannas da Silva, Lectures on Symplectic Geometry 2001 – Corr. 2nd printing (2008)

LECTURE NOTES IN MATHEMATICS

Edited by J.-M. Morel, F. Takens, B. Teissier, P.K. Maini

Editorial Policy (for the publication of monographs)

1. Lecture Notes aim to report new developments in all areas of mathematics and their applications - quickly, informally and at a high level. Mathematical texts analysing new developments in modelling and numerical simulation are welcome.

   Monograph manuscripts should be reasonably self-contained and rounded off. Thus they may, and often will, present not only results of the author but also related work by other people. They may be based on specialised lecture courses. Furthermore, the manuscripts should provide sufficient motivation, examples and applications. This clearly distinguishes Lecture Notes from journal articles or technical reports which normally are very concise. Articles intended for a journal but too long to be accepted by most journals, usually do not have this "lecture notes" character. For similar reasons it is unusual for doctoral theses to be accepted for the Lecture Notes series, though habilitation theses may be appropriate.

2. Manuscripts should be submitted either to Springer's mathematics editorial in Heidelberg, or to one of the series editors. In general, manuscripts will be sent out to 2 external referees for evaluation. If a decision cannot yet be reached on the basis of the first 2 reports, further referees may be contacted: The author will be informed of this. A final decision to publish can be made only on the basis of the complete manuscript, however a refereeing process leading to a preliminary decision can be based on a pre-final or incomplete manuscript. The strict minimum amount of material that will be considered should include a detailed outline describing the planned contents of each chapter, a bibliography and several sample chapters.

   Authors should be aware that incomplete or insufficiently close to final manuscripts almost always result in longer refereeing times and nevertheless unclear referees' recommendations, making further refereeing of a final draft necessary.

   Authors should also be aware that parallel submission of their manuscript to another publisher while under consideration for LNM will in general lead to immediate rejection.

3. Manuscripts should in general be submitted in English. Final manuscripts should contain at least 100 pages of mathematical text and should always include

   - a table of contents;
   - an informative introduction, with adequate motivation and perhaps some historical remarks: it should be accessible to a reader not intimately familiar with the topic treated;
   - a subject index: as a rule this is genuinely helpful for the reader.

   For evaluation purposes, manuscripts may be submitted in print or electronic form, in the latter case preferably as pdf- or zipped ps-files. Lecture Notes volumes are, as a rule, printed digitally from the authors' files. To ensure best results, authors are asked to use the LaTeX2e style files available from Springer's web-server at:

   ftp://ftp.springer.de/pub/tex/latex/svmonot1/ (for monographs).

Additional technical instructions, if necessary, are available on request from: lnm@springer.com.

4. Careful preparation of the manuscripts will help keep production time short besides ensuring satisfactory appearance of the finished book in print and online. After acceptance of the manuscript authors will be asked to prepare the final LaTeX source files (and also the corresponding dvi-, pdf- or zipped ps-file) together with the final printout made from these files. The LaTeX source files are essential for producing the full-text online version of the book (see www.springerlink.com/content/110312 for the existing online volumes of LNM).

The actual production of a Lecture Notes volume takes approximately 12 weeks.

5. Authors receive a total of 50 free copies of their volume, but no royalties. They are entitled to a discount of 33.3% on the price of Springer books purchased for their personal use, if ordering directly from Springer.

6. Commitment to publish is made by letter of intent rather than by signing a formal contract. Springer-Verlag secures the copyright for each volume. Authors are free to reuse material contained in their LNM volumes in later publications: a brief written (or e-mail) request for formal permission is sufficient.

Addresses:
Professor J.-M. Morel, CMLA,
École Normale Supérieure de Cachan,
61 Avenue du Président Wilson, 94235 Cachan Cedex, France
E-mail: Jean-Michel.Morel@cmla.ens-cachan.fr

Professor F. Takens, Mathematisch Instituut,
Rijksuniversiteit Groningen, Postbus 800,
9700 AV Groningen, The Netherlands
E-mail: F.Takens@math.rug.nl

Professor B. Teissier, Institut Mathématique de Jussieu,
UMR 7586 du CNRS, Équipe "Géométrie et Dynamique",
175 rue du Chevaleret
75013 Paris, France
E-mail: teissier@math.jussieu.fr

For the "Mathematical Biosciences Subseries" of LNM:

Professor P.K. Maini, Center for Mathematical Biology,
Mathematical Institute, 24-29 St Giles,
Oxford OX1 3LP, UK
E-mail: maini@maths.ox.ac.uk

Springer, Mathematics Editorial I, Tiergartenstr. 17
69121 Heidelberg, Germany,
Tel.: +49 (6221) 487-8259
Fax: +49 (6221) 4876-8259
E-mail: lnm@springer.com